Time and Measure

The Study of Time

Founding Editor

Julius T. Fraser†

VOLUME 18

The titles published in this series are listed at *brill.com/stim*

Time and Measure

Edited by

Stephanie Nelson
Arkadiusz Misztal
Walter Schweidler

BRILL

LEIDEN | BOSTON

Cover illustration: *The Propagation of Uncertainty*, 2020, three-channel video installation, 5 min. 40 sec. Video still. Photo © Emily DiCarlo (edited).

The Library of Congress Cataloging-in-Publication Data is available online at https://catalog.loc.gov

Typeface for the Latin, Greek, and Cyrillic scripts: "Brill". See and download: brill.com/brill-typeface.

ISSN 0170-9704
ISBN 978-90-04-72362-7 (hardback)
ISBN 978-90-04-72363-4 (e-book)
DOI 10.1163/9789004723634

This book is printed on acid-free paper and produced in a sustainable manner.

To the memory of Frederick Turner
dear friend and long-time ISST colleague,

a poet of Earth:

And so the poets' work is little more
Than cracking out the spirit they inherit
In the tall silver towers of poetry ...

and of Mars:

So Beatrice,
Taking into her hands her garden tools—
A dream of a Campanian burial,
A trope of lightness, and a wild new world—
Begins the cultivation of the void

∵

Contents

Figures

Notes on Contributors

Kerstin Cuhls
(Ph.D. 1966) is Scientific Project Manager at the Fraunhofer Institute for Systems and Innovation Research (ISI) in Karlsruhe, Germany, and Professor of Japanology at the Center of East Asian Studies at the University of Heidelberg. Her research is in foresight and time.

Emily DiCarlo
is an independent artist and scholar whose interdisciplinary practice focuses on temporality, site and collaboration. Her research has been supported by the Canada Council for the Arts and the Social Sciences and Humanities Research Council of Canada (SSHRC).

Ralph Gutknecht
(Ph.D. 1990) is Scientific Project Manager at the Fraunhofer Institute for Systems and Innovation Research (ISI) in Karlsruhe, Germany. With a background in Psychology, his research focuses on foresight and the interaction between humans and technology.

Paul A. Harris
is Professor of English at Loyola Marymount University. He served as President of the ISST from 2004–2013 and is co-editor of *SubStance* journal. Current projects include complementary books tentatively titled *Stone Dialogues* and *A Life Written in Stone*.

Karen Heald
is Professor of Interdisciplinary Art Practice at Wrexham University, UK. Her film work explores concepts of time, site, and creativity, from a painterly perspective. She collaborates broadly and disseminates her research globally via artist residencies, exhibiting in gallery contexts and non-traditional spaces, film festivals, conferences, and writing for publications.

Thomas Kantermann
(Ph.D. 1979) is Professor of Health Psychology at FOM University of Applied Sciences (Germany). He is Director of the Institute for Labour & Personnel and Research Academy (FOM). His research focuses on human chronobiology and sleep.

Ritsuko Matsumura
is a biologist studying molecular mechanisms of the mammalian circadian clock. She earned her Ph.D. in Agronomy from Kyushu University and has been serving as an Assistant Professor at Yamaguchi University's Research Institute for Time Studies.

Arkadiusz Misztal
is Professor in American Studies at Gdańsk University. He has published work on contemporary fiction, narrative theory, and the philosophy of time, including *Time in Variance* (2021), co-edited with Paul Harris and Jo Alyson Parker.

Friederike Mork-Antony
a hospital psychologist, has worked for several years as a research assistant, focusing on circadian rhythms and sleep disorders.

Stephanie Nelson
(Ph.D. 1992) is Professor in the Department of Classical Studies and the Core Curriculum at Boston University. She teaches widely in Greek and Latin literature and the Classical tradition and has published on Hesiod and Virgil, Ancient Greek comedy and tragedy, and the modernist reception of Classics, including *Time and Identity in Ulysses and the Odyssey* (2022).

Jo Alyson Parker
is Professor Emerita of English, Saint Joseph's University. Her publications include *Narrative Form and Chaos Theory in Sterne, Proust, Woolf, and Faulkner* as well as essays dealing with narrative and time. She has co-edited three volumes in The Study of Time series.

Chloe Garcia Roberts
is the deputy editor of the *Harvard Review* and a lecturer of poetry at MIT. She is also a translator from the Spanish and Chinese. Her latest book is *Fire Eater: A Translator's Theology* (2024).

Lanei Rodemeyer
works in the areas of phenomenology, continental philosophy, the philosophy of time, and feminist/gender philosophy of the body. She has published two books: *Intersubjective Temporality* (2006) and *Lou Sullivan Diaries (1970–1980) and Theories of Sexual Embodiment* (2018).

Walter Schweidler
has held chairs of philosophy at the universities of Dortmund, Bochum and Eichstätt. He has published on Wittgenstein, Heidegger, and time. Recent publications include *Wiedergeburt* and *The Other Time. Philosophical Approaches to the Past that has Never been Present.*

Sanyogita Singh
has a Ph.D. degree from Panjab University, Chandigarh, India. Her work discusses the dynamic materiality of time and the active temporality of matter within fiction.

Raji Steineck
Dr. phil. (Bonn University), is professor of Japanology at University of Zurich and previous president of ISST (2013–2023). He publishes on cultural philosophy and on Japanese intellectual history. His latest book is *Zen Time: Dōgen's Uji in Context* (2025).

Daniela Tan
teaches Japanese literature and religions at the University of Zurich. She has published extensively on contemporary Japanese literature, including an annotated translation of Kajiwara Shōzen's menstruation treatise for *Medieval Female Health in East Asia in Asian Medicine* (2025).

Fredric Turner
is a poet, a cultural critic, a playwright, a philosopher of science, an interdisciplinary scholar, an aesthetician, an essayist, and a translator. He is an author of numerous publications that span fields from poetry and literature to neuroscience and time.

Xiaoshi Wei
Ph.D. (Indiana University), is a research associate at SOAS, University of London, specializing in historical sound collections of China and the Turkic world. He directs the China Database for Traditional Music in Beijing.

Introduction

Arkadiusz Misztal, Stephanie Nelson and Walter Schweidler

This volume presents selected work from "Time and Measure," the 18th conference of the International Society for the Study of Time (ISST). The conference, whose usual triennial character was somewhat altered by Covid, was held from July 2–7, 2023, in Yamaguchi Japan, with the gracious and generous collaboration and assistance of the Japanese Society for Time Studies and, in particular, Makoto Ichikawa, the President of JSTS, and Kenta Fujisawa. the Director of the Research Institute for Time Studies at Yamaguchi University. Over sixty participants from Japan, China, Europe, the Americas, Australia, and India participated in the panels, which were held in one plenary session during the day, while the evenings were dedicated to keynote speeches. Here Ritsuko Matsumura gave the Founder's Lecture, "Molecular Genetic Study of the Circadian Clock," published here, while Jan Zalasiewicz spoke on the technosphere as the new measure of change in the Anthropocene, and Michelle Bastian on the role of citizen scientists in recording the shifting ecosystems now replacing our once regular seasons. The topics presented in the conference overall ranged from the measure of time in music, to "clockwork kings" in literature and comics, to the crucial role of measuring time in law, economics, and medieval Japanese medicine, and generated widespread and intense interdisciplinary conversations. Particularly notable was Ricardo Uribe's "Indigenous Clockmakers, Schedules, and Quantitative Time in the Spanish Colonies (16th Century)," winner of the Founder's Prize for New Scholars, given in honor of J. T. Fraser.[1] Thanks to the Japanese Society for Time Studies, sponsors Seiko and Casio, financial support from the Yamaguchi Tourist Association, and the indefatigable work of Ms. Hiroko Hirata and ISST's own Daniela Tan , the papers were augmented by an amazing day of visits to the Akama Jingu Shrine, Karato fish market, the castle town of Chofu, and the Akiyoshidai limestone park, pictured here in Karen Heald's essay. It would be hard to find a more vivid experience of the radical differences in measures of time than that experienced in the ninth-century shrine, the bustling modern fish market, and the uplifted reef of Paleozoic limestone and its Pleistocene fossils.

The conference participants reflected again on time as measured in multiple aspects and through multiple media, in the art installations that ran

1 An extended version of this paper was subsequently published in *KronoScope: Journal for the Study of Time*, see Uribe (2024).

 | DOI:10.1163/9789004723634_002

continuously during the week of the conference, under the acute guidance of Emily DiCarlo. These ranged from the two described in this volume to John Steck Jr's cyanotypes of the decay of flowers or Andrew Cozzens' surprisingly moving videos of passers-by holding their breath for as long as they could—a very visceral reminder of how unthought-of time-measures inform our experience. Finally, and worth particular note, the considerable number and range of presentations on measuring time in religion, theater, tradition, music, and the visual arts, and in particular, Nahoko Suzuki's paper on the Noh drama, *Atsumori*, were further enhanced by the remarkable privilege of a Noh performance of the well-known piece, *Funa Benkei*, at the Noda Shrine in Yamaguchi City, featuring renowned Noh actor, Yoshitaro Imamura. We also had the honor of dinner at the traditional Japanese restaurant and museum, Saikotei, and of the company of some of the highly skilled actors from the ensemble, who spoke about their art, illustrating traditional Japanese stage motions, both tragic and comic. Even further, Yoshitaro Imamura graciously accompanied us on our visit to the Akama Jingu Shrine, and treated us to another, impromptu, recitation of the famous final battle of Dannoura, which occurred in 1185 in just the place we then were. To hear the traditional verse in its own living environs meant also experiencing for ourselves the stilling and enchanting of time through the beauty of the human voice.

The theme of the conference, Time and Measure, reflects J. T. Fraser's fundamental observation that time exists through a nested hierarchy of *umwelten*, worlds as they are experienced by, and so measured by, each one's particular inhabitants. These range from the first, atemporal level to the purely probabilistic time of particle-waves through to the biotemporal, then the teleological, intentional time of the nootemporal, and finally (whether seen as an element of the nootemporal or as a separate level), the sociotemporal time of organized groups of humans. In the essays of *Time and Time Again*, for example, Fraser draws a contrast between this crucial insight and Plato's fundamental opposition of Being and Becoming. While Plato privileges the atemporality of Being, and implies a single uniform ontological measure for time in its essential nature as Becoming, Fraser's view sees radically different and conflicting levels nested within one another, each developed from and independent of those below, and each with its own unique measure of time. And as these different measures range from that of particle-waves to the humanly shaped ecosystem we now inhabit, it is natural that they appear throughout our experience, as mirrored in the multiple disciplines of the ISST. As individual papers explored the move from Newtonian to quantum physics, the geological record, chronobiology, technologies and colonization, and social phenomena such as time experienced in prison or during the Covid lockdown, they brought out as well, in fine detail, the intricacy of Fraser's levels of *umwelten*, each both dependent on and in conflict with those beneath it. Perhaps most crucially, however, we see the importance of Fraser's insight that measures of time are not merely different in scale, but crucially different in kind. As the variety of topics considered at the conference reveals, this is as true of the difference between deep time and the continual acceleration of the Anthropocene as it is of the different measures of time in a countdown, or legal "futures," or the Japanese "Memorial Day of Time," or in T. S. Eliot's words, in "music heard so deeply / That it is not hard at all, but you are the music / While the music lasts" (1995, 199). In this way the variety of topics considered in the conference and in this volume itself reveals the depth and importance of Fraser's theory of time.

The initial section in the collection, "Time's Past and Future," directly concerns the temporalities of Fraser's Extended Umwelt Principle, with both Raji Steineck and Lanei Rodemeyer addressing Fraser's theories by considering further implications and possible new directions for his hierarchy of levels. Steineck points out the difficulty, acknowledged by Fraser, that his temporal levels both evolve into the development of time and are also described as themselves occurring in time, even at levels where there is no stable relation of before and after. In response he suggests, in contrast to Fraser, that the Extended Umwelt Principle be taken in its original sense, as an epistemological rather than an ontological approach to time. In arguing for a view of time measured not by number but by relation of attributes Steineck sets up as well an interesting conversation with Turner, both internally in his own chapter and in dialogue with the essay by Turner that ends this section. Steineck also points out that the move he suggests, from Fraser's ontological Generalized Umwelt Principle back to his heuristic Extended Umwelt Principle, would restore the distinction that Fraser deliberately removed, between reality and appearance, a distinction, Steineck argues, which we now need more than ever.

Citing the same problem of the nested hierarchy as both in time and creating time, Lanei Rodemeyer takes Fraser's theory in another direction by putting it into conversation with the phenomenological approach of Edmund Husserl. While Husserl, naturally, has nothing to add to the atemporal, prototemporal, and eotemporal levels of Fraser's theory, the levels before the advent of human perception, Rodemeyer suggests that using Husserl's levels of constitution as a way to think through the higher levels of Fraser's hierarchy creates a more nuanced view. Thus, Husserl's "hyletic flow," the basic, subconscious flow of sensory input that underlies our perceptions of things as meaningful objects, bridges Fraser's biotemporal and nootemporal levels. Introducing it into Fraser's hierarchy thus has the advantage of acknowledging different levels within the vast range of Fraser's biotemporal sphere, ranging as it does from the simplest life forms to the human. Going even further, Rodemeyer suggests adding a new level to Fraser's hierarchy, the associo-temporal, lying between the bio- and nootemporal. The bottom of this level would correspond to Husserl's Hyletic Flow, and the top to his Passive Synthesis (subjective relations of present, past, future), while Husserl's next level, of Active Constitution (subjective time as object) would correspond to Fraser's nootemporal level. This view has the advantage of introducing a place for the cognition of the higher non-human animals as beyond simple instinct, but not at the level of conceptual thought. A similar nuance is introduced as two levels of Husserl's scheme, Interpersonal Intersubjectivity (objective time) and Intersubjective Community (abstract objective time) further extend Fraser's nootemporal/sociotemporal level(s), allowing for an expansion in Fraser's model of social consciousness.

Moving from Fraser specifically to other philosophic considerations of the measure of time, Walter Schweidler examines how Leibniz's work helping design the great garden at Herrenhausen Castle led to the conceptual groundwork for the monad, as it mediates between totality and infinite continuity. Bringing together Steineck's relation of epistemology and ontology with the interest in our relation to the natural environment seen in the volume's second section, "Ourselves as the Measure of Time," Schweidler begins with Leibniz's paradoxical declaration that "each branch of every plant, each member of any animal, each drop of its liquid parts is also some such garden or pond." He then explores Leibniz's idea of the monad, each of which mirrors the whole, by looking first at the ancient Roman idea of each individual's particular "genius" and then at that idea's successor, the medieval angel. Finally, he finds that the crucial element Leibniz discovered in the garden at Herrenhausen is time, which, like the deviation in angles that make the garden in fact not rectangular, is perceived imperceptibly. Just as in viewing stars what we "see" occurred millions of light-years ago, so an ordinary tree-branch that I see today contains also, imperceptibly, the tree from which it came, and the tree which generated that tree, back through to the origin of time. In this harmonious compatibility of totality and infinity, time, Lebniz argues, is to be conceived as rooted in the unfolding of the world-whole.

Paul Harris then takes our consideration of "Time's Past" even further in looking at the deep time of a specific stone, an amphibolite which he discovered in the Cache de la Poudre River in the Rocky Mountains of Colorado. As Misztal will connect walking as a measure of time to its embodied experience, and as Schweidler connects Leibniz's understanding of imperceptibles to his experience at Herrenhausen, so Harris interrelates the geological origins of the rock, likely formed at a depth of tens of kilometers nearly 1.7 billion years ago, to his own physical experience of it. Connecting also to Heald and DiCarlo's accounts of their art installations, with which the volume will end, Harris also discusses his attempt to convey the stone's time to others by placing it on view, a process which critically involves its now human-created setting and positioning. Finally, again reflecting "Ourselves as the Measure of Time," Harris' account of his own physical and meditative experience with the stone also suggests a possibility of the human reaching back to the stone's time, just as the stone has been thrust into ours. The essay thus suggests a form of geologic timefulness through which one can integrate large temporal scales with the measure of a stone's time.

In contrast to Harris' reflection on slow time, and to Misztal's account of the pace, which will open the next section, Frederick Turner closes this section by looking at the constant acceleration and metamorphosis of time, not simply as we experience it in the world around us, but also conceptually, in the move from principles to emergent systems. While Schweidler's account of Leibniz located us in the Newtonian world of natural law, Turner contrasts Newtonian physics to the quantum, while also moving from a consideration of the past which time itself may have had to a consideration of time's future. Turner thus argues that the idea of a fixed universal unit of measurement, or measuring-stick, has now given way to measurement through a new form of comparison, a comparison of systems mutually affecting each other and so mutually altering. Suggesting a term, *logogenesis*, for the continual emergence of new rules, principles, and forms of measurement that emerge with and regulate this new order, Turner draws out the implication of the packed-ness of current time, creating a measure not by abstract units but by events, a view again in interesting conversation with Steineck's opening piece. Turner then illustrates this measure through the immense complexity that arises in the "turnings" of the computational "travelling-salesman" problem, as well as in the ordinary experience of city vs highway driving. His final paradoxical conclusion is that when compared with space, largely empty as it is, it is our own human time that is immense. The thought is beautifully captured in the passage from Turner's epic poem, *Genesis* that concludes the piece, comparing the richness of poetic tradition on earth to the bare first beginnings of such another tradition on Mars. The passage thus leads as well into the next section, on time's measure as precisely a reflection of our own experience.

Opening the next section, "Ourselves as the Measure of Time," Arkadiusz Misztal continues Rodemeyer's phenomenological approach by looking at the embodiment of time in the pace, with particular reference to the thinking of

Maurice Merleau-Ponty and Henry David Thoreau. The phenomenological nature of time now appears more viscerally, as Misztal describes how walking, as an embodied measure of time, involves an inherent engagement with the environment of the walker. This engagement, more experienced than thought, further connects external conditions such as weather, time of day, and terrain to the internal, as the walker's heartbeat, breath, and balance continually adjust to the surroundings. And yet, as we see in Misztal's exploration of Merleau-Ponty and Thoreau, the experience is also a deep form of thinking. In this way walking both sets thinking in motion and, in transcending self-engendered mobility, opens the mind to thoughts that are not so much self-generated as revelatory and "possessive," a "peripatetic thinking" experienced as ideas and thoughts incarnating into the rhythmically moving body.

The next two papers of this section, on human chronobiology, move to a biological consideration of ourselves as the measure of time. As Heald will mention, many participants in the conference were experiencing the effects of travel from the Americas or Europe to Japan, making the topic of the human biological clock particularly pertinent. Kerstin Cuhls' reminder of the crucial social importance of taking chronobiology seriously, and, on an individual level, of identifying one's own particular chronotype, thus resonated deeply, as she reported on the findings of CIRCADIA, a research project on the social effects of sleep loss due to screen-technology use, particularly as seen during the Covid lockdown. Similarly, Ritsuko Matsumura's Founder's Lecture, explaining the important research of her group on the molecular constitution of chronobiology, had very immediate resonances at the time. Together the two papers not only examine the measurement of time as an inherent element of our biological existence, but also focus on the individual particularity of our biological clocks, as they continually readjust in their almost, but not quite, 24-hour circadian rhythm. And as these readjustments are done largely in response to external light, the papers raise further questions concerning the interdependence of our individual biological clocks with both the natural world, in its succession of day and night, and with our own human-created world of light-emitting devices.

Moving from our innate biological measure of time to our measure of time through culture, Daniela Tan and Xiaoshi Wei explore, respectively, cultural time as measured through the body in medieval Japanese legends, and through the *sybyzghy*, a wind instrument used to accompany the tales, legends, and histories that measure and maintain Kazakh culture over time. Both papers look at how oral traditions measure time through narrative, and at how time also alters narrative, as traditional legends both shift and persist in their cultural significance. Tan thus examines the bending of biological and narrative time in Japanese legend, seen in a theme of elderly women suddenly regaining their menstrual flow upon entering religiously prohibited areas. She argues that although an increasingly patriarchal society saw the reappearance of menstruation as a sign of ritual impurity, this distorts an earlier pre-10th century view, that saw the female resistance to prohibition as leading rather to a return of vital forces. Wei, similarly, looks at the embeddedness of legend in culture in examining how *sybyzghy* performances manifest the porous boundaries between epic, legendary, and historical measures of time in the cultural history of the Kazakhs. Working from a focus on the indexicality of time, Wei considers not only the critical importance of context in these performances, but also, in a fascinating connection to both Tan and Misztal, the embodiment that the instrument itself involves. In exploring the intimate connection of a culture's measurement of time with their traditional narratives, the two pieces also give a new resonance to the imagining of the newly created time of a to-be-created culture with which Turner's piece ends.

The next and final section, "Time's Uncertain Measure," begins with two prose poems by Chloe Garcia Roberts, "Temporal Saturation" and "Whens." As the pieces consider the multiplicity of timeframes within each of us, they reflect the preceding section, but as Garcia Roberts' rhythmic prose also captures how ever-shifting moments and patterns of time shape our experience, they resonate with the studies of the uncertainty of measuring time which follow. Further, as Garcia Roberts points to "certain people in the life of each of us, figures ringed in mothglow, who can be expressed as refrains of richest temporal saturation," she explores as well how our measures of time are determined through our interrelatedness as humans, also seen in both Singh's and Parker's studies.

We then move to two pieces of literary analysis, Sanyogita Singh's consideration of Graham Swift's *Waterland* and Jo Alyson Parker's study of Ted Chiang's "Anxiety is the Dizziness of Freedom." As Singh shows, *Waterland*'s non-linear narrative time resonates with Deleuze's account of repetition and difference, showing how the changes emerging through repetition also reveal the past as continually reshaped by present and future. Revealing again the interdependent times of culture, narrative, and the individual, this resonance permeates not only the family history, the historical awareness, and the overall consciousness of the internal narrator, Tom Crick, but also the very language that constitutes the novel. It also appears deeply in the landscape that informs Crick's awareness and the novel as a whole, the River Ouse and the Fens. Singh's piece thus brings out once more the dependence of human time on the lived environment in which it is experienced, a theme that resonates with Wei's account of how the Kazakh steppes inform their cultural history, and goes back to the various approaches to this theme in Schweidler, Harris, and Misztal.

Moving again, though now in a quite different register, from the measure of time as it connects us to the past to the measure of time as it connects us to the future, Parker examines the idea of branching timelines in Chiang's novella. Here, as in *Waterland*, a movement away from a unitary, linear view of time also brings the nature of the self into question, but now in asking what might result from an ability to measure oneself against the "paraself" of an alternative future. With a fascinating reflection of Turner's introduction of quantum physics, but now focused on many-worlds theory, Parker goes on to explore the psychological and philosophical implications of continually multiplying reality, now through Chiang's positing of an ability to examine multiple "unled lives."

We conclude both the section and the volume with two artists reflecting on work presented at the conference, Karen Heald on her moving image, *The Timekeeper and the Hour Glass* and Emily DiCarlo on her *Propagation of Uncertainty*, a three-channel video installation filmed at the National Research Council (NRC) in Ottawa. Both installations juxtapose a multiplicity of time-measures, while also returning us to the theme of embodiment, as Heald works with the Butoh dancer, Angelina Kornecka, at the Welsh medieval fortress, Chirk Castle, and as DiCarlo sets her own body against the time lab in which she is filmed. Both artists also use their papers to set their work in a broader philosophic context. Heald thus discusses how her tripartite conception of "in-between-ness," "being-in-between" and "transitory strata" grows from Julia Kristeva's work on intertextuality and the semiotic *chora*, as well as Hito Steyerl's work on "groundlessness" and the "poor image." Her consideration of the breakdown of any separation between observer and observed resonates both with Turner's piece and Singh's, while her consideration of time as consciousness recalls Garcia Roberts' verbal recreations of time. Setting *The Timekeeper and the Hour Glass* in the context of her past work, Heald also considers the role of context in time, particularly as reawakened by her return to Japan after a space of 17 years.

Heald's piece leads, finally, into DiCarlo's, who sets her own embodied time against an institutional striving for an exact and scientific measurement of time. As an explicit study in the measurement of time, DiCarlo touches on themes that have run throughout the volume: the "post-real time" she describes as used to determine Coordinated Universal Time resonates with Steineck's reflection on appearance and reality; her observation that the process means also that "in reality, a multiplicity of deviated pasts informs our future's official, singular present" recalls Schweidler's study of Leibniz; her presentation of the human body as measurement returns to Misztal's study of pace, as well as the entire second section of the volume; and her account of the constant acceleration of current time reflects Turner's essay. Similarly, DiCarlo's account of the need to continually reset mechanized time, and of Siffre's experiences in resetting his own circadian rhythm, is given further depth and detail by the chapters contributed by Cuhls and Matsumura. Above all, however, DiCarlo reflects on the multiple timeframes created by our embodiment and embedding in the global temporal infrastructures of late capitalism. Her study thus brings out as well the increasing divide between lived time and the time of critical computational processes. In the installation itself, the visual and audial process of standardizing Coordinated Universal Time, which serves as background to DiCarlo's gradually revealed body, is also an ideal last image for the existence, juxtaposition, and conflict of different measures of time. DiCarlo's exploration of the infinitely complex layering of Coordinated Universal Time, and her addition of "research-creation" as an alternative, embodied methodology, thus reflects not only Rodemeyer's expansion on Fraser's nested hierarchy, but also joins it in revealing the depth and importance of Fraser's insight.

1 Time's Past and Future

While the three sections of the present volume span the varied disciplines implicated in an understanding of time, like Fraser's work they also reflect a fundamental ontological investigation, which seems to merit a closing consideration of their interrelation. The title of the first section, "Time's Past and Future" points to a theme that runs through the first five articles, connecting the measure of time with reflections on the ontological and epistemological relation between the measuring subject and the measured object. As the crossing point between the two, and connecting this section to the next, "Ourselves as the Measure of Time," we find the human body in its complex engagement with the world to which it belongs and yet also incorporates. A paradigmatic approach to this crossing point leads us back to what Raji Steineck calls the apparent paradox of the beginning of time, a paradox perhaps defused by the "understanding that temporal order *precedes* its measure in countable mass units." The paradox also makes it clear, however, that the question of "time and measure" is not at all restricted to the aspect of time concerned in "applied" science or technological practice, or even everyday temporal order; it is rather a question which leads us into the speculative center of the meaning of temporality for our understanding of existence.

2 Ourselves as the Measure of Time

As we move from this ontological question to the second section we see as well that our own existence, i.e. our relation to ourselves, is a paradigm for our understanding of temporality. As time always, from its most archaic characterizations until today, has been represented in relation to the existence of beings who "have" it—as "lying" before or behind them, as "going" fast or slow, as "turning," "urging," "fleeing" etc.—, this is much more than a mere metaphorical illustration; it is again a hint of time's ontological essence. As time is intertwined with those who experience it, and as there is no "becoming" without or beyond time, then temporality itself seems in some way to be "passing" by or even "aging." That means time has a past. Time is "had" in the sense that it remains itself only in the passing from what it was to what it is. It is the infinite process of catching up with itself. It is, in the words of Emmanuel Lévinas, the "astonishing divergence of the identical from itself" (1993, 28) or, as Maurice Merleau-Ponty puts it, "the affecting of self by self" (2005, 494). This is what we see through ourselves, when we, here and now, turn back to time's way into this present moment.

3 Time's Uncertain Measure

As our reflection on ourselves as the measure of time also reveals the essential relation between time and that side of subjectivity that, in our everyday constitution, remains unconscious to ourselves, it also connects to the theme of the third section of the volume, "Time's Uncertain Measure." The philosophic aim of making the imperceptible perceivable may be seen as key to the understanding of the paradox that time was time before it consisted of a sequence of units measured in numbers, years, hours, or seconds. This understanding that time is older, and extends later, than whatever can be measured by it points as well to the necessary uncertainty of measures of time. Perhaps the classical phenomenological characterization of the flux of time, Husserl's "living presence" ("lebendige Gegenwart"), is appropriate for the expression of this insight. When we measure time, we are not watching from a certain distance or from outside what is going on "in" the course of temporal existence; we are not, in the words of Frederick Turner, "applying a measuring stick to a theoretically ascertainable quantity." Rather, we are acting from "inside," in some way "doing" time. We express our experience of the living presence of the world to which we belong, i.e. our own living and continually changing presence, as we also become aware of the relation in which any living being stands to its ancestors and its heirs. The measure of past and future must then be uncertain, since past and future are not simply earlier and later chapters in a book that we are presently reading. They are rather, as Leibniz once said, born out of the process which makes the present being "pregnant" with what is to come. Just as time itself is in its own way "aging," our experience of it is shaped by our natural desire to keep up with its ever-changing acceleration, to stay ahead of itself, to create the meaning of what enables us to be not threatened, but rather freed, by the measure of time.

References

Eliot, T. S. 1995. *Four Quartets*. London: Faber and Faber.

Fraser, J. T. 2007. *Time and Time Again: Reports from a Boundary of the Universe*. Leiden, The Netherlands: Brill Academic Publishers.

Lévinas, Emmanuel. 1993. *Otherwise than Being, or Beyond Essence*. Translated by Alphonso Lingis. Berlin: Springer.

Merleau-Ponty, Maurice. 2005. *Phenomenology of Perception*. Translated by Donald A. Landes. London: Routledge.

Uribe, Ricardo. 2024. "Indigenous Clockmakers, Schedules, and Quantitative Time in the Spanish Colonies (16th Century)." *KronoScope* 23: 45–176.

PART 1

Time's Past and Future

CHAPTER 1

Toward a Wider Horizon: Restricting the Umwelt Principle

Raji C. Steineck

Abstract

J. T. Fraser's "Extended Umwelt Principle" (below: EUP) takes its clue from work by biologist Jakob von Uexküll on the sensory and motor faculties of animals, most famously the tick. Uexküll argued that the world as it is accessible to an animal is defined by the spectrum and reach of its sensory-perceptive faculties and its abilities for locomotion. Together, they define what can appear as part of the "umwelt" of this animal. Fraser applied this idea beyond the realm of animals, both in the direction of inanimate matter, and in the direction of beings with higher cognitive faculties and their associations. His further movement from the EUP to a "generalized umwelt principle" then posited that the reality of any given being is defined by the causal and, in case of living beings, perceptive/cognitive nexus it has with the world. I argue that there are two problems with this extension: first, on the material side, it reduces the range of causalities that may actually have effects on a given entity. Second, on the conceptual side, it ignores the distinction between "umwelt" and "reality." I propose a return to the EUP as a heuristic device to formulate the temporal structures *constitutive* at a given level of reality, while avoiding the naturalist fallacies that burden the later versions of Fraser's theory. In this revised version, level-specific temporalities remain important as a measure of the "umwelt" of an entity; they define the temporal structures that must remain in place if this entity is to preserve its integrity. At the same time, the revised EUP upholds crucial distinction between reality and appearance.

Keywords

Extended Umwelt Principle – J. T. Fraser – ontology of time – epistemology of time – Ernst Cassirer

•••

 | DOI:10.1163/9789004723634_003

> [A] horizon ... wider than any fact which we could ever discover in relation to the world we belong to.
>
> WALTER SCHWEIDLER, *The Other Time*, 77

∵

The "Extended Umwelt Principle" (below: EUP) is a key concept in J. T. Fraser's later work and the basis for his time-based ontology. According to this principle, the internal temporal and causal structure of any given segment of reality is best described in terms of the stimuli that can be received and the effects that may be caused by its inhabitants (whether they be photons in the realm of light or nation states in the human world). In thus conceiving of "species-specific universes" (TTA,[1] 154), not only for humans, animals, and plants, but also for stones, electrons and photons, Fraser built a "hierarchical theory of time" (TTA, 41)—a theory that explains the evolution of temporal structures of growing complexity, from the a-temporal realm of pure energy to the convoluted world of human societies, where dreams of eternity clash with drives for instant gratification.

For all its merits in guiding the theoretical imagination, the EUP is nonetheless problematic, at least in the holistic interpretation that Fraser gave to it. In his interpretation, the principle "identif[ies] epistemology (whatever we may be capable of knowing) with ontology (whatever we may assume to be the case)" (TTA, 155). To claim this, however, is to expand an originally heuristic device into a metaphysical hypothesis. To distinguish this metaphysical, holistic version of the principle from the heuristic, epistemological one, we can call it with Fraser the "generalized umwelt principle" (Fraser 1999, 25; abbreviated in the following as GUP). Not only did Fraser not give sufficient reasons for expanding the "extended" into the "generalized" umwelt principle, he also had to admit that the expansion led to a contradiction in his evolutionary theory of time (TTA, 156). I argue that this contradiction can be resolved if we stay with the EUP (the epistemological principle), but refrain from ontologizing it into the GUP. In doing so, I propose to retain the essential "distinction between 'appearance' and 'reality', between what things seem to be and what they are" (TTA, 38, quoting Bertrand Russell) that Fraser was proud to "remove" (ibid.). In other words, I endorse the heuristic merits of the EUP and with them, the insights of Fraser's "hierarchical theory of time" concerning

1 Most references to Fraser in this article are to essays collected in his 2007 volume *Time and Time Again* (2007), abbreviated here and in the following as TTA.

the specific temporal umwelten of each level. In contrast to Fraser, however, I continue to distinguish between these umwelten and "reality." The reality of a certain domain or level of nature is only incompletely represented in the umwelten it comprises. As a consequence, I aver the existence of an 'arrow of time' present before the emergence of life—which resolves the contradiction noted by Fraser. The pertinent discussion will further lead me to the question of whether time is essentially defined by "number," as in Aristotle's famous statement that it is "just this—number of motion in respect of 'before' and 'after'" ("The Internet Classics Archive | Physics by Aristotle" n.d., book II.11). Here I argue that typology is more fundamental than metrics: time is first and foremost defined by a relation between attributes, with number (and topology) coming second.

1 Evolution without Direction?

According to the hierarchical theory of time, certain features central to the everyday notion of time—most importantly the notion of time's inexorable unidirectional progression, its "petty pace from day to day," are latecomers to the universe: without life, all processes are inherently reversible, and there is therefore no fixed direction of time.

Fraser identified three integrative levels of nature that are without a stable order of past, present, and future: first, the *atemporal* level of photons "in which no meaning can be given to the idea of 'event,' to conditions of 'before-after'" (TTA, 155); second, the *prototemporal* level of quantum events, which "may be identified but their temporal positions never precisely determined" (TTA, 156); and third, the *eotemporal* world of lifeless, massive matter, where "time may be said to 'flow' yet it does not have a 'direction'" (TTA, 156). He further insisted that in a universe where all elements belong to these three levels of photons, particles, and massive matter, there is no "preferred direction" (TTA, 46) of time, no order of past, present, and future (cf. TTA, 156): "By the Umwelt principle, we must [...] admit that [these] temporalities or causations are not only appropriate but also sufficient—that is complete—for the processes and objects considered" (TTA 41), and therefore:

> [T]he temporal umwelts of the physical world, though without a now or directed time, must be regarded as complete in themselves, even if from the point of view of our daily experience of time they appear incomplete and terribly odd. The experience and idea of time's passage must be brought to physics; they cannot be derived from it.
>
> FRASER 1999, 35

A contradiction arises, however, when physics is thus equated with ontology, as occurs in the GUP. This is evident from the accounts Fraser gave of the first stages in the evolution of time.

> The oldest, stable integrative level of nature is the chaotic universe of electromagnetic radiation. Out of that primeval chaos *arose* objects that had nonzero rest mass and traveled at speeds less than that of light; these particle-waves *came to constitute* the second stable integrative level of the world. A billion years later—more or less—massive matter *began* to freeze out, *eventually forming* the 10^{10} galaxies of the universe. These islands of matter float in an immensity of almost complete emptiness. The massive matter of the galaxies-chemical elements and their compounds in different abundance-constitute the third stable integrative level of the world.
>
> TTA, 41–42; my emphases; see also Fraser 1999, 27

I have marked by italics the expressions in this account that are incompatible with the application of the GUP. In the umwelten of radiation and wave-particles, nothing can be said to "arise" or "come to constitute," because that implies a stable relation of before and after. According to the GUP, however, such a relation does not exist at this level of nature. Similarly, it is impossible to speak of "beginning" and "eventually forming" in the umwelt of massive matter, since both expressions refer to an irreversible direction in time. Fraser noted this "self-contradiction" in his ontology, but attributed it to a fault of language, saying that "we have no vocabulary in which the evolution of time itself may be discussed in self-evident terms" (TTA, 155).

The contradiction can be resolved, however, if we accept that even in the early stages of the history of our universe, its *reality* did not coincide with the sum total of the temporal umwelts of its constituents. When seen in this way, Fraser's description simply illustrates the fact that the reality of the inorganic physical universe cannot be fully described in terms of the umwelten it comprises. Their structures are a necessary condition for the evolutionary events in question, but the very fact of these events also shows that the reality of the universe may contain aspects that are lost to the umwelten present in it. This is not simply a question of an ex-post description from a human point of view, a narrative that anthropomorphizes the universe. It is, or so I argue, a realistic account, and one that is in line with the well-considered premises of Fraser's theory.

Consider the following: if we accept Fraser's proposition of a nested hierarchy of integrated levels with level-specific causalities and temporalities, we must accept that there is temporal directionality *before* the emergence of biotemporality, simply because the notion of a nested hierarchy in itself implies a directed sequence. The reason is that the higher temporalities can only exist on the basis of the lower ones: there is no nootemporality without biotemporality, and no eotemporality without prototemporality (cf. TTA, 43). Therefore, once the level of prototemporality has emerged from atemporality, there is an irreversible sequence, revealing temporal direction as a feature of the universe/reality. Put differently, once the second level of the "nested hierarchy" has come into existence, there is already an "arrow of time."

2 Time before Its Measure

Notably, the temporal sequence implied in the nested hierarchy is not based on measurement: it only requires that we can identify the existence of two discernibly different states—in this case: A_p and E_p, absence or existence of particles with a rest mass larger than zero –, where one state presupposes the other and can therefore only come into existence *after* the other has been established. This leads us to a fundamental insight concerning the nature of time: measurable "units" belong to its secondary, or higher order, features. While time may be described as the order of change, this is an order that precedes measure, taken in terms of cardinal numbers counting "units" of time.

The first numbers we can attach to time are the ordinal numbers which indicate position in a sequence: first this, second that, third something else. First atemporality, second prototemporality, third eotemporality. But before they may be thus counted, they must be identified; there must be at least two discernibly different states within the same universe. This establishes the first order of time: this happened, and at another time, that. This is still a purely typological order in which different times are distinguished, without a rule connecting them. Direction comes into play only when "that" is understood to be somehow based on "this." From there, it becomes possible to count: "first this, then (second) that," establishing a sequence of 'before' and 'after', and thereby, a direction of time. Units of time, which may be counted by numerical measure, presuppose a plurality of recurring sequences of events of type A, B, C (countable as first, second, third iteration of A, B, C, and so forth) in a stable relation to each other (3 iterations of A during one iteration of B *and* 15 iterations of A and 5 iterations of B during one iteration of C).

In terms of the hierarchical theory of time, because the existence of radiation is understood as a condition for the emergence of particle-waves, and the existence of particle-waves as a condition for the existence of massive matter, we can say that irreversible sequence is a temporal feature of the universe *before* the temporal umwelt of biotemporality arises, even if irreversibility was not a feature of any of the "umwelts" of its constituents. Once a second and third level of temporality have emerged from out of the initial, chaotic, a-temporal *umwelt*, the universe has acquired a history: the arrow of time has emerged, even though there is not yet an *umwelt* that incorporates this fact. Notably, it even makes sense to apply a numerical measure to this history, as Fraser did in saying "[a] billion years later—more or less—massive matter *began* to freeze out, *eventually forming* the 10^{10} galaxies of the universe." We can accept this way of speaking in terms of "years" because the regularities already in place in the world of particle-waves (which we understand to behave with certain periodicities) can be meaningfully related to the regularities established later, after planet earth formed, circling the sun. But understanding that temporal order *precedes* its measure in countable mass units may still help to defuse an apparent paradox about the beginning of time: it is possible and logically consistent to say that time has a beginning, and that something was *before* the beginning of time, without being able to calculate that *before* in terms of years, hours, or seconds. The initial state before time, we may say, was in itself timeless (a-temporal), without events, duration, or sequence. It became part of a sequence only ex post, with the advent of prototemporal umwelten and, thereby, temporal sequence.

3 An Umwelt of the Universe?

As mentioned above, my proposition is that an arrow of time exists before the emergence of life and biotemporality, a proposition that also implies that the reality of the universe comprises more than what is accounted for in the sum total of its umwelten. I will expand on the reasoning behind this below. Before doing so, however, let me discuss a position that could integrate the above insight into the GUP's fusion of epistemology and ontology, of "knowing" and "being." This position suggests that we understand the evolution of temporal sequence as an umwelt not addressed, but possibly implied by Fraser: the umwelt of the universe, or nature, itself. This appears to be the position of, for example, Fred Turner, who made the following comment on the presentation that formed the basis of this paper:

> [It] seems to me that the radical point of Fraser's vision is that this distinction [between "knowing" and "being"] itself is an invalid imposition on the *real* reality, in which the state of knowledge of a system is a fundamental aspect of its reality—as evidenced by decades of quantum

> experimentation. "Knowers-about" are not, pace Cassirer, confined to human beings. [...] Light is both the fundamental material of which the universe is made, and the fundamental form of information. Thus the history of the universe is simultaneously a history of the combining of photons into more and more complex and determinate forms of energy and matter, and what one might call the working out of a computational (informational) process. It's an onto-epistemological universe, a "state of being" but also a "state of knowledge" in the quantum sense. But it is incomplete, and thus evolving thermodynamically and computing itself into a more comprehensive state of knowledge. It is its own Laplace Calculator, tasked with establishing its own complete account of itself and thus always incomplete as its state of knowledge enlarges.
>
> TURNER 2023

Turner's prose is, as always, powerful and suggestive—and one may find pleasure and consolation in the thought that our human knowledge is actually part and parcel of the self-exploratory process of the universe. This is one way to transcend, conceptually at least, the boundaries of our mortal existence.

I am, however, skeptical about extending the notions of "umwelt" and "knowledge" in this fashion, and for two reasons. First, as noted by Fraser, the universe is a singular and comprehensive entity: there is no time and space apart from the universe. What, then, is its umwelt if it does not have an outside? It may be possible to conceive of an internal umwelt—as dreams or propriosensations may be seen as part of the umwelt of a living being, for example. But does it make sense to conceive of an entirely internal umwelt—which implies something that exclusively receives and acts on information from parts of itself? And can one say that the universe (as a holistic entity) has propriosensation, that it "perceives" and "reacts to" events within itself? What physical structures would both separate the whole from its parts and connect the parts back to the whole?[2]

Second, I believe that Turner's use of the terms "information" and "knowledge" is evocative, but also equivocal. It is admittedly possible to define 'information' in a general manner that includes processes in quantum physics, by identifying it with "the value currently attached or instantiated to a characteristic or variable returned by a function, f(x), or a process" (Losee 1997, 259). Notably, however, this definition distinguishes the information from the process, its input and output:

2 On these structural requirements for implementing a "self," see Plessner 2019.

> [I]nformation is not the process itself. The input to the process is not information about the process, although it clearly may be information about another process. The output is also not information by itself; the values in the output are information only in the sense that they are information about the process and the input, that is, information in the context of the process and its input.
>
> LOSEE 1997, 260

A process, e.g. radiating light, may carry information, but it is not itself information. The output of the process likewise becomes information not simply because it triggers an effect, but only when parts of it are received as "values" revealing something "about" the process and its input.

One may want to define 'cognition' in a similarly general manner as the process of (correctly) evaluating information (that is, referring "values" to what they are "about"), and to define 'knowledge' as the result of such evaluation. One could then say that a bird 'knows' a worm when it evaluates signals emanating from a wriggling, elongated object and starts picking at it. One could further build a meaningful notion of 'machine knowledge', when a machine discriminates between 'noise' and 'information' in what it receives from its sensors. In doing so, however, it would make sense (and be only consistent when interpreting Fraser) to apply the extended umwelt principle. As a consequence, one would need to distinguish between different hierarchically nested levels of knowledge and cognition. This would serve to prevent the inadvertent imputing of characteristics of human knowledge and cognition to the "bionoesis" of birds and worms, or the "eonoesis" of machines or, possibly, mountains. I leave it to those knowledgeable about information processing on these lower levels to work out the details and turn to human cognition to make my case.

4 Transcending the Umwelt: Symbolic Representation

Fraser defined the "organizational level of nature" that evolved together with homo sapiens as "the processes and structures created by the human mind, using its skills for the symbolic transformation of experience and its capacity to appreciate non-present objects and events" (TTA, 42). Along with the Neo-Kantian philosopher Ernst Cassirer (mentioned by Turner above), I submit that the latter capacity is a function of using recursive systems of representative symbols—language being the most important of these. I also suggest that such a use of representation creates a fissure in the bubble of the umwelt, where the notion of 'reality' creeps in and in so doing addresses the question posed above concerning a possible reality beyond that of particular umwelten.

'Symbols' are higher-order signs, whether either material objects or events created to convey meaning. Smoke is a sign of fire. Smoke signals are symbols: they are created to convey an idea of something other than smoke or fire, for example, the presence and numbers of an enemy or prey. Symbols convey meaning in different ways: *expression* uses the sensuous characteristics of the artifact to indicate analogous content—think of a dismissive gesture, or a rising, uplifting melody. In *representation*, the meaning is one step removed from the symbol's shape and regulated by a code. A good code uses analogies of shape and so supports representation with expression—"sorrow" in English, German, and Japanese is indicated by word stems with dark vowels. But expression here is not required: French "tristesse" isn't less sad because of its bright vowels.

Expression "presents" what it is meant to convey (Cassirer 1927, 301–2), but by the same token ties it to the actual symbol and the instance of communication. That is not the case for representation. Representative symbols can point to something entirely absent to the senses, and can convey the same meaning in a different situation. They are also far more flexible: one symbol can replace another—as Shakespeare's Juliet says, "that which we call a rose by any other name would smell as sweet" (*Romeo and Juliet*, Act II.2, quoted after Shakespeare 2003, 960). We can substitute a word or phrase or paint a different picture to convey the intended meaning. We can even change the medium of representation and move from language to image and vice versa. This hugely broadens the horizon of possible references. But it also brings in the question of adequacy: does the symbol (or chain of symbols) convey "fact" or "fiction"? Representative meaning is always "posited" meaning, and Cassirer identifies this ability to "posit" (*setzen*) something as the "new dimension" of meaning that arises with language (or more broadly, with representation) (Cassirer 1927, 302). What is posited, however, can also be wrongly posited, be it out of error or the will to deceive.

Cassirer may be jumping to conclusions when he believes that, as a consequence, representation necessarily leads to the question of "being" (*Sein*, ibid.). The study of higher animals shows that recognition of error and deceit in the use of representative signs, e.g. those indicating a specific kind of danger, may lead to only correction and punishment (Hauser 1992)—and not to an inquiry into the quality of the signs themselves. But it is true that representation enables such inquiry, and even to some extent calls for it. What Cassirer specifically seems to have in mind are systems of representative symbols that allow for recursion, symbols that can represent other symbols. This option becomes available with representation, but seems so far to be exploited only by human beings. Where symbols stand for other symbols, representation can point to something that is not only absent to the senses, but that can never appear to them, such as a perfectly straight line, or the meaning of the word "reason." Such abstraction reinforces the drive for examination: Does "squaring the circle" really "make sense"? Could a winged horse really exist? And further, since representation allows for the indication of types and all-quantifiers: Does anything really exist, or is everything just an illusion?

5 Looking beyond the Umwelt

The first two questions pierce the bubble of our umwelt; the third one opens a wide rift. This rift, the question of how appearances relate to reality, continues to haunt humanity (witness Russell); but it has also set it free to explore and examine whatever is presented to the human mind by the senses, symbols, symbol-based instruments, and any combination of both—as is beautifully represented in the Flammarion image of a man looking beyond the firmament (Figure 1.1).

To illustrate the decisive difference that comes with questioning reality, when, in Cassirer's words, the question of being is being posed, consider a palm tree on one of the tropical lagoons in the pacific. The palm tree is a living being; as such it has its sensors and effectors. Responsive to conditions of light and shade, wind and water, it turns its leaves or grows its roots in a certain direction. We can therefore rightfully impute to it some form of "biocognition." Currently, our palm tree is affected by global warming. Some elements of this are part of its *umwelt*: it may grow faster because of the increased level of carbon dioxide, or be weakened and start to wither as an effect of increased groundwater salinization (Beringer et al. 2023; Gomes and Prado 2007). At some point it will be submerged in the ocean, which rises because of the ice melting at the poles (Garcin et al. 2016). The rising sea levels that lead to salinated groundwater may or may not yet be a part of its umwelt. Combustion engines, filling stations, refineries, petroleum tankers, and oil fields in faraway countries contribute to global warming (Ghosh and Ghosh 2020), but our palm tree has no sensors informing it of their existence: they are no part of its umwelt. Oil companies, lobbying groups, fossil-friendly political parties, and governments protecting big oil interests are a further important part of its problem (Lantushenko and Schellhorn 2023), that does not appear in its umwelt. Nevertheless, these factors at some point determine its fate. "Reality bites," as a film title from the 1990s put it (Hawke and Ryder 2003). And reality is more than appearances. "Reality" is the proper term for everything that is the case in and possibly even beyond the universe. It is independent of the

FIGURE 1.1 Beyond the Firmament. Engraving. In *L'atmosphère: météorologie populaire*, by Camille Flammarion, Paris: Hachette, Paris 1888, 163
IMAGE IN PUBLIC DOMAIN. WIKIPEDIA, ACCESSED MARCH 28, 2025. HTTPS://EN.WIKIPEDIA.ORG/WIKI/FLAMMARION_ENGRAVING#/MEDIA/FILE:FLAMMARION.JPG

question of "appearances," and is posited as their ground and as the measure of their adequacy. The palm tree has no means to form such a concept of reality; it cannot question appearances, it cannot fathom what is beyond the reach of its sensors. That is why it is enclosed in its *umwelt* and, in the words of Uexküll and Fraser, lives in it like in "a soap bubble" (TTA, 39). In contrast, as explained above, humans have learned to use symbols that are not what they represent. These symbols make it possible to think of what is absent, even inaccessible to the extended sensors of the human umwelt. They may be deceptive. But precisely because of that, they allow us to form a notion of reality that transcends appearances and serves as their critical measure. We've witnessed over the past years what happens when that distinction gets lost, when people get trapped in "soap bubbles" of social media that only feed them what fits their cognitive apparatus.

6 Conclusion

Human cognition, operating through recursive systems of representative symbols, is enabled to formulate questions that call the umwelt itself into question. That is, precisely, the function of notions like "reality" or "truth." They challenge the soap bubbles created by the senses and received views. They also create a fundamental insecurity, a conflict between the wish to be reassured and the awareness that there is no final reassurance, because, in the words of the medieval Zen master Dōgen, "when one side is illuminated, the other side is in the dark" (Tanahashi 1999, 35). That, in turn, leads to a never-ending demand for more knowledge and for knowledge that is more secure. Humanity has developed two main strategies to cope with demand: it has created notions of a "perfect intelligence," omniscient and removed from all doubt, accessible to enlightened vision or through the testimony of prophets and holy scriptures. It has also followed the less shining path of patiently expanding its knowledge by critically investigating appearances and synthesizing the results in systems under constant revision. Empiricists may find it hard to accept that the latter endeavor, the endeavor of the sciences and humanities, is, in the final analysis, driven by an idea—and one that can never be fully accounted for in experience. I suggest they consider how this idea cracks all cognitive bubbles and drives forward the open-ended exploration of experiences and what may lie behind or beyond them.

The EUP is a powerful device to this end, particularly when understood as a heuristic, epistemological principle. I have shown that restricting its application in this manner resolves a self-contradiction in Fraser's hierarchical theory of time and explains how an arrow of time may have emerged in the universe before the advent of life. It thus leaves us, paradoxically, with more rather than less.

References

Beringer, Tim, Christoph Müller, Julia Chatterton, Michal Kulak, Sibyll Schaphoff, and Yvonne Jans. 2023. "CO2 Fertilization Effect May Balance Climate Change Impacts on Oil Palm Cultivation." *Environmental Research Letters* 18 (5): 054019. https://doi.org/10.1088/1748-9326/accbd5.

Cassirer, Ernst. 1927. "Das Symbolproblem und seine Stellung im System der Philosophie." *Zeitschrift für Ästhetik und allgemeine Kunstwissenschaft* 21: 295–322.

Flammarion, Camille. 1888. "Un missionnaire du moyen âge raconte qu'il avait trouvé le point où le ciel et la terre se touchent," in: *L' atmosphère météorologie populaire*, Paris: Hachette, p. 163. https://www.e-pics.ethz.ch/index/ETHBIB.AD/ETHBIB.AD-v1.0_26975.html.

Fraser, J. T. 1999. *Time, Conflict, and Human Values*. Urbana: University of Illinois Press.

Fraser, J. T. 2007. *Time and Time Again: Reports From a Boundary of the Universe*. Brill.

Garcin, Manuel, Myriam Vendé-Leclerc, Pierre Maurizot, Gonéri Le Cozannet, Bernard Robineau, and Alexandre Nicolae-Lerma. 2016. "Lagoon Islets as Indicators of Recent Environmental Changes in the South Pacific—The New Caledonian Example." *Continental Shelf Research* 122 (July): 120–40. https://doi.org/10.1016/j.csr.2016.03.025.

Ghosh, Swapan K, and Bablu K Ghosh. 2020. "Fossil Fuel Consumption Trend and Global Warming Scenario: Energy Overview." *Global Journal of Engineering Sciences* 5 (2): 1–5.

Gomes, Fábio P., and Carlos H. B. A. Prado. 2007. "Ecophysiology of Coconut Palm under Water Stress." *Brazilian Journal of Plant Physiology* 19 (December): 377–91. https://doi.org/10.1590/S1677-04202007000400008.

Hauser, M. D. 1992. "Costs of Deception: Cheaters Are Punished in Rhesus Monkeys (Macaca Mulatta)." *Proceedings of the National Academy of Sciences* 89 (24): 12137–39. https://doi.org/10.1073/pnas.89.24.12137.

Hawke, Ethan, and Winona Ryder. 2003. *Reality Bites: USA 1994*. Los Angeles: Universal Studios.

Lantushenko, Viktoriya, and Carolin Schellhorn. 2023. "The Rising Risks of Fossil Fuel Lobbying." *Global Finance Journal* 56 (May): 100829. https://doi.org/10.1016/j.gfj.2023.100829.

Losee, Robert M. 1997. "A Discipline Independent Definition of Information." *Journal of the American Society for Information Science* 48 (3): 254–69. https://doi.org/10.1002/(SICI)1097-4571(199703)48:3<254::AID-ASI6>3.0.CO;2-W.

Plessner, Helmuth. 2019. *Levels of Organic Life and the Human: An Introduction to Philosophical Anthropology*. Fordham University Press.

Schweidler, Walter. 2023. *The Other Time: Philosophical Approaches to the Past That Has Never Been Present*. Verlag Karl Alber. https://doi.org/10.5771/9783495994115.

Shakespeare, William. 2003. *The Complete Works of William Shakespeare*. Essen: Magnus Verlag.

Tanahashi, Kazuaki. 1999. *Enlightenment Unfolds: The Essential Teachings of Zen Master Dogen*. Boston, London: Shambhala.

"The Internet Classics Archive | *Physics* by Aristotle." n.d. Accessed December 12, 2023. http://classics.mit.edu/Aristotle/physics.4.iv.html.

Turner, Frederick. 2023. "EUP," E-Mail to the author, July 8, 2023.

CHAPTER 2

Temporality in Layers: a Conversation between J. T. Fraser and E. Husserl

Lanei Rodemeyer

Abstract

Despite their obvious differences, J. T. Fraser and Edmund Husserl also have approaches to temporality and temporal experience that offer parallel touching points, suggesting that a comparison of the two might be fruitful. As an approach to such a comparison, this paper first works through the different levels of temporal measure discussed by each of these theorists. I then consider how these two hierarchies might align with one another, and whether they might work together to produce a more detailed and nuanced system. Finally, I carry out two analyses intended to bring these two positions further into conversation and which suggest that introducing Husserl's concepts and leveled analyses into Fraser's "nested hierarchy" might allow for new approaches to our understanding of time in general.

Keywords

Edmund Husserl – J. T. Fraser – temporality – time – levels

One[1] would think that J. T. Fraser, who employed his "nested hierarchy" to understand the natural world, and Edmund Husserl, often accused of idealism in his phenomenology, would have very little in common. Several facts counter this presumption, however: Both scholars recognized temporality as an essential component for describing the world; both pointed to human existence as a key feature in what we understand as temporality; both presented various

1 A shorter version of this article was first presented at the International Society for the Study of Time conference, in Yamaguchi, Japan, in 2023. I thank the organizers and attendees for their support of this project. Further, I am grateful to the editors of this volume, and especially Stephanie Nelson, for their careful feedback and editorial comments, bringing this article to a new "level" of clarity.

 | DOI:10.1163/9789004723634_004

temporalities based upon "levels"; and, interestingly, both identified "intentionality" as crucial to human engagement with the world. While this does not mean that Fraser and Husserl were making the same arguments, it does suggest that a conversation between them could be productive. This paper intends to begin that conversation.

First, I work through the ways in which each of these theorists presented different levels of temporal measure.[2] For Fraser, the "Extended Umwelt Principle" describes a "nested hierarchy" of temporalities in the natural world, from electro-magnetic radiation (lowest) to human social existence (highest). For Husserl, I argue, embodied subjects experience the world through similar levels of temporal constitution, from a flow of raw sensory material (lowest) to communal or abstract social existence (highest). As it will turn out, these two systems are complementary to a surprising degree.

1 Fraser's "Nested Hierarchy" of Temporalities in the Extended Umwelt Principle

The "lowest" level of temporality for Fraser is what he calls *atemporality*, which is not really temporality at all. This is the presumed pure chaos that existed prior to the formation of the universe, or as Fraser puts it, "Atemporality does not stand for nonexistence but for a world of absolute chaos, a total absence of causation, such as is believed to have existed in the primeval universe at the instant of the big bang, or as it exists in a black hole" (Fraser 2007, 47). For Fraser, the atemporal level corresponds to that primeval chaos when everything was traveling at the speed of light—although we should remember that this is prior to the existence of any *thing*—and where nothing like causation could exist. Here we have energies without organization, direction, or causal relation.

From this chaos emerged particle-waves, which have only probabilistic causal relations. Fraser refers to this level as *prototemporality*, because causal and temporal relations at this stage are not yet determined or specific, and

2 Given that most readers of this article may have little to no familiarity with either J. T. Fraser's or Edmund Husserl's work, the first two sections of this paper are primarily reviews of some main facets of their work. I will first describe Fraser's "nested hierarchy" of temporalities, which he describes in multiple writings. In the following section, I will lay out Husserl's "levels of constitution," which I find to be quite consistent throughout his work even if he did not describe them as explicitly as I do here. The following sections will contain more analytical work that examines some of the results of placing these two theories in dialogue with one another.

yet lay the groundwork for the concrete temporal relations that arise in the next level. As Fraser puts it: "Out of that primeval chaos arose objects that had nonzero rest mass and traveled at speeds less than the speed of light; these particle-waves came to constitute the second stable integrative level of the world" (Fraser 2007, 42). As we do not yet have causality, we also do not yet have *time* as we commonly understand it; rather, we have the beginning of temporality as evidenced in the motion and speed of particle-waves. The relations of these particle-waves can only be statistically probable, such as the statistical probability we find in the movement of electrons or in radiation, making this level one of probable speed and motion as opposed to temporal causation.

Bodies of great mass, galaxies, and other objects take form out of this prototemporal existence, bringing us to Fraser's next level, *eotemporality*. At this level, we (finally) find deterministic causation, where specific objects relate to one another in determinable ways based upon their mass, speed, direction, etc. Fraser calls this level *eotemporality*, which he defines as the "oldest of developing forms" (2007, 46), based upon the root "eo," as in Eos, goddess of the dawn (2007, 156). At this level, we find both succession and the deterministic causal relations of physical objects. In this sense we finally have a sense of time: we can employ the time "t" that we find in the calculations of physics, and we can understand basic temporal relations in a universe filled with planets, galaxies, and so on. However, the succession here is without any necessary direction. In other words, succession at this stage does not necessarily move from a "past" into a "future," but rather, is simply an ordering without an innately preferred direction. In a similar way, we can follow the order of numbers either "forward" or "backward" without there being any "natural" direction to the numbers themselves. As Fraser puts it, "Eotemporal events are countable and orderable, as are the natural numbers and, as do those numbers, they form a pure succession without a preferred direction" (2007, 46). Time is, at this level, simply a "constant" or a "variable" by virtue of which we can recognize and calculate physical change with more specificity.

For Fraser, many theorists of time focus primarily on the eotemporal level, ignoring or dismissing the levels to come, and sometimes even forgetting about the prior levels of atemporality and prototemporality. On the basis of the eotemporal level, though, biological life is able to arise, but—and this is important—*only* when certain conditions are met *at the right time*. In other words, the next level, which Fraser calls the *biotemporal* level, is not only the level where biological life emerges, but more importantly, it is where *simultaneity and process become evident and essential.* As Fraser points out,

"Biochemical events that should happen simultaneously must, those that should not ought not, or else the integrity of the life process will be lost. Instant by instant synchronization in cooperative functioning, governed by stable principles, assures collective viability" (2007, 44). And later, he describes the biotemporal world in this way: "The characteristic connectivity among events in the biotemporal world [...] is organic intentionality directed toward short term, concrete goals, serving the continuity of the life of the organism" (2007, 46). In other words, life does not just need particular components (certain chemicals, temperatures, etc.) but also events (collision, electrical surge, etc.) at precise and often intersecting, simultaneous moments, in order to survive; if those components and events are not present at the required moment, then life ends—or it doesn't begin. For this reason, unlike the prior temporal levels which had no "now," at the biotemporal level, the "now" is an absolutely necessary component that enables the life within it. "The crucial break in the development of temporalities came with biogenesis; specifically, with the need of organisms to provide for continuous, internal synchronization of biochemical processes" (Fraser 2007, 156).

Further, in addition to a now-moment where all components must be present, *life requires a temporal direction*, the direction of the process that follows its beginning. A process must follow a specific order in time, and because these developments rely upon temporal direction, the biotemporal level is where this temporal direction emerges. Thus, life requires both a present and a temporal direction—unlike the prior temporal levels where neither a present "now" nor temporal direction were necessary. In this way, according to Fraser, temporality itself evolved—it evolved with the manifestation of life. Temporality becomes more complex, and the now and temporal direction emerge with the life that requires it.

Biological life leads further, however, to intentional, thinking life. Through the development of symbolic thought and language, according to Fraser, a thinking individual is able to recognize itself as having a past, living in a present, and moving into a future. Thus, a new evolution of temporality emerges. Thinking life brings about the level of *nootemporality*, where human beings experience themselves in time, and where they engage with their existence through goals and projects as well as through reflection upon memories and past events. "The characteristic connectivity among events of the nootemporal world [...] is that of intentionality, directed toward concrete or symbolic goals, serving the continued integrity of the self" (2007, 46). Here we find time in its fullest sense and as we usually understand it: encompassing a past, present, and future, as well as a directionality from past events to future possibilities.

Depending upon which of Fraser's texts you are referencing, the next and final level is either understood as a development within the nootemporal world,[3] or as another level altogether.[4] This level is the *sociotemporal*, which arises from organized groups of people with shared goals and interests, but which is grounded in the temporal awareness of the individuals who make it up (just as each of the temporal levels is "nested" in those that are prior to it). I would argue that this level stands on its own, rather than being included in the nootemporal level, for the following reasons. First, social groups often engage in causal relations well beyond what could be derived from a sum of individual human beings. In other words, groups function according to their own rules and causal relations, such that they are not reducible to the individuals that make them up. Second, the temporal extension of societies and organized groups stretches well past any individual life, or even the lives of several generations of a family group. So, once again, we cannot simply find the components of sociotemporality in nootemporal individuals. Thus the sociotemporal level is a stage beyond the nootemporal level both in functioning according to different rules and causalities, and in having a time exceeding that of nootemporality. In fact, Fraser himself seems to recognize this in one of his descriptions: "[…] [T]here are reasons to believe that in dealing with societal time we shall need a new language whose vocabulary and syntax derive entirely from social and not from individual experience" (2007, 157; see also 158). For this reason—and because it aligns with the levels we find in Husserl's phenomenology—we will understand the sociotemporal level to be a separate temporal stage which emerges beyond nootemporality.

It is interesting to note that Fraser's differentiation between these temporal levels is based upon the distinct causal relations employed in each,[5] as well as (and relatedly) upon the different mathematical calculations required for each of the "lower" levels, and the meaningful or symbolic intentions guiding each of the "upper" levels. Thus the prototemporal world employs probability to understand the relations of its objects, while eotemporality is based instead in direct physical causation. These levels also each have their own characteristic physical or abstract objects which undergo a major shift in focus in the transition from one temporality to the next. For example, particle-waves in prototemporality give way to large mass objects, such as planets, in eotemporality. In other words, each "level" functions according to its own rules, which

3 See Fraser (2007, 44–45).
4 See Fraser (2007, 156–57).
5 See Fraser (2007, 43 and 157–58).

are directly related to its particular causality and objects.[6] We will discuss this further in our analyses, below, especially with regard to the transition from one level to the next. For now, we turn to a description of Husserl's "levels of constitution."

2 Husserl's Levels of Constitution,[7] and Their Relation to Temporality

Husserl's levels of constitution are based on analyses of (human) experience, and for this reason, when we compare Fraser with Husserl, Husserl's levels will fall primarily alongside the higher levels of Fraser. We should also keep in mind that Husserl's are levels of *experience* in contrast to Fraser's levels of *existence*. Nonetheless, despite this distinction, we will find some interesting correspondences. For now, let's look at what Husserl has to offer.

---

6 Here we already encounter a similarity with Husserl, who distinguished between "regions of being" in his earlier philosophical works, by pointing to the different mathematical, causal, or meaningful rules that guide the objects of each region. See, for example, Husserl (1983, 54), where he points out that, "I busy myself, let us say, with pure numbers and their laws: Nothing like that is present in the surrounding world, this world of 'real actuality'. [...] [D]uring such [arithmetical] busiedness single numbers of numerical formations will be at the focus of my regard [...]; but obviously this factual being-there-for-me [of the natural world], like the factually existent itself, is of a different sort." While Husserl is here arguing that the natural world and the arithmetical "world" require different "attitudes" for us to engage with them, this position is based upon his understanding that the different "worlds" function according to different rules and laws: for example, the natural world functions according to causal laws, whereas the arithmetical world does not have causal relations but rather numerical, formal, and/or formulaic relations. I am pointing out here that the distinctions between levels for Fraser, which are based upon differences in causality, mathematics, and intentionality, parallel Husserl's distinctions between regions based upon their different laws and rules. Nevertheless, there remain distinctions: Fraser's nested hierarchy is understood to have each higher level *emerge* from the one below it. The lower level provides not only the conditions of possibility for the next one, but is also the factual basis for the rising up of the subsequent, "higher" level. In contrast, Husserl's "regions of being" are understood—at least in his earlier work—to be relatively exclusive from one another; the laws and rules that distinguish one region from another also demonstrate that they cannot relate to one another. In this way one cannot emerge from another. Nevertheless, it is worth noting that both theorists found that the rules, causality, and/or objects of different "worlds" were what fundamentally distinguished them from one another. Further, please note that the "regions of being" discussed in this footnote are not the same as Husserl's levels of constitution discussed in this paper. The levels of constitution are levels of experience, rather than regions of objects.

7 These levels of constitution are not explicitly presented in Husserl's work, but rather, can be gleaned through extensive reading across his texts. I present them in another context in Rodemeyer (2021).

The "lowest" level for Husserl relates to what he calls the "*hyletic flow*," the accession of sensory material that grounds all our perceptual experiences. This is simply our basic flow of sensory input in embodied human existence. This sensory flow underlies all our perceptions of things as meaningful objects, grounding them with sensory content. So, for example, my visual perception of a tree outside is based in sensory moments of green and brown shades, and these color moments appear through a continual flow of visual input as I look at the tree. While I do not *see* "flowing green shades"—I instead see a tree—my consciousness uses these sensory data as the foundation for what I take to be a tree in my visual experience. But this flowing of sensory input requires *presence*. In other words, the sensory contents of our experience must be present to consciousness for them to be part of our perceptual experience.[8] Thus, this lowest level already reveals the basic workings of what Husserl calls, "inner time-consciousness," because it points to the necessary presence of our sensory flow. Inner time-consciousness is the activity of consciousness that takes our experiences and connects them in multiple ways, but especially in a temporal ordering. This temporal ordering establishes a past, present, and future, and creates the foundation for recollection and anticipation. In this way, inner time-consciousness is the source of our most basic experience of time, and while it functions at all levels of constitution, it is already evident in how the hyletic flow is taken up in an ordered way.

In addition to the temporal ordering that is carried out through inner time-consciousness, Husserl finds that similar types of sensory contents are here already associated with one another (green with green, for example, or sound with sound). In other words, the sensory flow is not just a blur or confused blending, but rather, it already has separate components that connect and occur together, in contrast to others. This indicates another basic aspect of temporality that is enabled through inner time-consciousness: duration. Groupings of sensory data within the flow require a structure that constitutes them beyond any immediate moment, so that they are recognized as some type of persistent unit of similarity. Simply put, they require a basic structure of duration. Thus, the contents of the hyletic flow arise through a temporal presence, appear together as homogenous, distinguished from heterogeneous content, and remain in their appearing through an extension into duration. As Husserl puts it:

8 This is not an argument about the existence of the material things that might be the sources for our sensory input. Rather, it is a logical argument concerning the sensory input that plays into our constitution of perceived objects.

> In every living-moment various kinds of temporal objectlike formations are constituted, each one has a momentary Now together with the momentary horizon of retentional having-been and the momentary protentional horizon. This momentary structure is continually in the process of streaming and is synthetically united, precisely thereby constituting this temporal object as identical with itself, e.g., a lasting sound.
>
> HUSSERL 2001, 172

Temporally speaking, then, the hyletic level offers us our most basic experiences of presence, simultaneity, and duration, due to the basic functions of inner time-consciousness.

While Husserl's lowest level enables these basic experiences, the next level, "*passive synthesis*," points to connections between the present and the past through a perceived similarity of whole events or objects. This level allows us to associate similar, yet distinct, objects and events with each other, saving us from the need to recollect, in every situation, how certain objects or events need to be understood or how they function. When I see my cat, for example, I do not need to recall all my former experiences with her in order to recognize her; rather, I recognize her because this present experience is couched in many past experiences that provide depth and connection to this perception of her right now. Passive synthesis provides us with temporal depth in its associations of similar objects and events by establishing a connection between past experiences and our present. It builds our connection to our past experiences, enabling us to move smoothly into our future. As Husserl says, "Between the noematic components of something present and something remembered we find a phenomenologically peculiar connective trait that can be expressed in the following way: Something present recalls something past" (Husserl 2001, 163).

For Husserl, passive synthesis is a function of consciousness that operates without any required attention on our part. Rather, the functioning of inner time-consciousness makes this passive synthesis of associations, connecting present to past (and future), possible: "We see very quickly that the phenomenology of association is, so to speak, a higher continuation of the doctrine of original time-constitution." (163) In other words, through association and passive synthesis (and based upon inner time-consciousness), we already have a connection to and relationship with the past, even if we do not experience explicit recollections. This is how we develop habits, recognition of familiar events and places, and so on.

The next level, "*active constitution*," is where we relate to meaningful objects directly and in our awareness. Here, objects present to me are constituted as wholes through their various profiles. At this level, we have a fully constituted experience of time as past, present, and future, and we understand the notion of time as a concept or meaning. In other words, we can understand time as an *object*—an object that is different from the inner time-consciousness that underlies our constitutive experiences: "We must carefully observe the distinction between this *phenomenological time*, this unitary form of all mental processes within *one* stream of mental processes [...] and the '*objective*', i.e., the *cosmic time*" (Husserl 1983, 192, emphasis in original). In addition, we have a more fully developed sense of simultaneity and duration, as whole objects endure and appear coexistent with one another. We also recognize the mutual influence of past, present, and future on one another, especially with regard to our interpretation of events, our perspectives on situations, and our involvement with events and objects. Simply put, this is the level where we, as temporal subjects, relate to the world of objects in time, although, as in Fraser's nested hierarchy, this relation obviously has its roots in the two prior levels.

But for Husserl, we attain a fuller sense of objectivity—objects as "there for everyone"—when we understand our world intersubjectively. Thus we turn to the first of *two* intersubjective levels, that of "*interpersonal intersubjectivity*." At the interpersonal intersubjective level, we attain a higher level of objectivity, as "there for everyone," where objects appear as fully transcendent, i.e., beyond my own consciousness of them. It is consequently at this level that we gain a sense of "nature" and the "natural," and further, where we gain a sense of time that underlies all of the natural world.

> In that way the *coexistence of my <polar> Ego and the other Ego*, of my whole concrete ego and his, my intentional life and his, my "realities" and his—in short, a *common time-form*—is primally instituted; and thus every primordial temporality automatically acquires the significance of being merely an original mode of appearance of Objective temporality to a particular subject. In this connexion we see that the temporal community of the constitutively interrelated monads is indissoluble, because it is tied up essentially with the constitution of *a world and a world time*.
>
> HUSSERL 1999, 128, emphasis in original

Here temporality is shared with those individuals around me, and when the present is experienced as shared, we begin to understand simultaneity and duration as objectively experienced "by everyone" as well. It is through this shared sense of time that we can make plans for the future, talk meaningfully about past events, and recognize mutual experiences in the present. Time is now beyond me, and beyond any individual; it is both the intersubjectively shared time that connects us and the objective time of the natural world.

The second level of intersubjectivity—and the last of Husserl's levels that we will discuss—is "*intersubjective community*," where time transcends even my interpersonal relations, and becomes formalized through clocks, temporal measurements, and scientific abstractions. Here we gain a sense of objective time and space that is well beyond that of any individual life experience. "Openly endless Nature itself then becomes a Nature that includes an open plurality of men [...], distributed one knows not how in infinite space, as subjects of possible intercommunion" (Husserl 1999, 130, emphasis in original). Time becomes an abstraction, beyond my life, beyond even the life of my specific community. It encompasses communities beyond my own, and further, it includes intergenerational histories and global events. Here time is a universal concept that includes not only the past and future of all communities of subjects, but also the past and future of objective nature. This is also the basis for time understood scientifically, i.e., the time of objective measurement. It is on this level, as Husserl says, that we see that "there *can exist only a single community of monads*, the community of *all* co-existing monads. Hence there can exist *only one Objective world*, only one Objective time, only one Objective space, only one Objective Nature" (Husserl 1999, 140, emphasis in original). We can now apply this objective sense of time to the physical world, and communicate about it through symbolic language within the scientific community. While objective time was already achieved at the level of interpersonal intersubjectivity, at this level, time doesn't become "more" objective, but rather, the formality, universality, and infinity of time become more apparent, and abstract developments of time (such as the mathematical formulas of physics, or the digital clock) can be developed.

These levels of constitution are the ways through which Husserl sees us as experiencing the world, and, accordingly, they point to different aspects of our experiences. However, although I have described them as sequential, in a way similar to Fraser's "nested hierarchy" of temporalities, these levels are always functioning in our experience, and, more importantly, their influence goes in *both* directions rather than merely building upward, as we see with Fraser. In other words, the hyletic flow of my sensory experience feeds upward into how I experience objects and other people, but conversely, the meanings developed in societies or between small groups of people can also filter down to affect how some sensory data might be more prominent than other data. The meaning of a specific object might be affected either by a change in sensory data or by a shift in intersubjective interpretation. Similarly, my habitual relations to an object can influence how I relate to a similar object, just as, in the other direction, a new objective meaning could alter my habitual engagements. In this way Fraser's and Husserl's "hierarchies" function in different ways and elucidate different aspects of the world and our experiences of it. Nevertheless, given that both recognize distinctive layers to temporality itself, it seems worthwhile to consider how they might work in conjunction with one another. We thus turn now to a comparison between Fraser's and Husserl's levels of temporality, where we find some intriguing similarities and differences.

3 Phenomenological Constitution within the Extended Umwelt

I would first like to sketch out how these two sets of "levels" might line up alongside each other before presenting a diagram. Starting at the "bottom" of the two sets of levels, Fraser's eo-, proto-, and atemporal levels—as purely natural, probabilistic, mathematical, and/or hypothetical—would not correspond with any of Husserl's phenomenological levels, which concern human experience.

When we get to Husserl's notion of the hyletic flow, though, things become much more interesting. Because it is so focused on sensory experience, the hyletic flow could be understood as splitting itself between two of Fraser's levels, biotemporality and nootemporality. As our sensory connection to the natural world, hyletic flow links up with biotemporality, but as part of our conscious experiences, the sensory, it aligns rather with nootemporality. In fact, Husserl often struggled with different "levels" within the hyletic flow—sometimes referring to an ultimate "primordial hyle" (Urhyle) or pure hyletic flow that would underlie the hyletic flow as it contains primordial associations. For this reason, I suggest that Husserl's "hyletic flow" may be at both Fraser's bio- and nootemporal levels, reflecting Husserl's description of the sensory body as a "turning point" between the sensing individual and the natural world.[9] In fact, we will later find that this overlap could lead to some interesting insights about the levels of temporality.

Similarly, Husserl's two levels of passive synthesis and active constitution might both correspond to Fraser's nootemporality. Here we see in Husserl two different levels of the thinking person: direct and self-aware relations with the world in our awareness, on the one hand, and relations of similarity and association in the background of our thought, on the other.

Finally, I suggest the following: both of Husserl's two levels of interpersonal intersubjectivity and intersubjective community align with Fraser's sociotemporality. So, where Fraser tentatively suggested a social level beyond the individual thinking person, Husserl identified two levels, first an interpersonal

9 See Husserl (1989, 167–69).

relation between individuals and in small groups and then, on the highest level, a more abstract sense of community, generations, and history.

In this way, Fraser's approach to time through nature and the history of cosmology offers insights and developments, especially when it comes to our understanding of objective time, that we will not find in Husserl, just as Husserl offers a more detailed breakdown of levels of time as humanly experienced. In fact, given Fraser's naturalistic approach that nevertheless appreciates human perception, and Husserl's development of subjective experience while also recognizing the importance of the natural world, we could argue that their projects combine into a deeper overall picture, informative for an analysis of time overall.

Given this, I offer a visual outline of how these two approaches line up with one another:

Fraser	**Husserl**
Atemporal	–
Prototemporal	–
Eotemporal	–
Biotemporal	Hyletic flow (presence, duration, and succession)
Nootemporal	Hyletic flow (presence, duration, and succession)
Nootemporal	Passive synthesis (subjective relations of present, past, future)
Nootemporal	Active constitution (subjective time as object)
Sociotemporal	Interpersonal intersubjectivity (objective time)
Sociotemporal	Intersubjective community (abstract objective time)

As becomes even clearer on this chart, Husserl offers more differentiation amongst the higher levels, which relate to our human experience of time, while Fraser describes several levels "beneath" the "intentional" ones, delving into the cosmological aspects of time. Both recognize the importance of both human experience and the natural world, and both look to ways in which we can integrate these two, seemingly disparate, views of temporality, the subjective, human view, and the objective view of nature—without subsuming either one to the other.[10] Given this, it appears promising to consider the

10 This appears to be a focus for several theorists around the same time period (Husserl [1859–1938] and Fraser [1923–2010] overlapping somewhat in their lifetimes). Another contemporary philosopher for whom this was an important issue, and for whom time

positions of these two theorists more carefully in light of one another to see how this combined scaffolding of time might develop. Given my own experience as a Husserl scholar, I offer two analyses from the perspective of Husserl's works, intended to further develop our understanding of time through a dialogue with Fraser.

4 Analysis A: the Influences between the Levels

As just discussed, given that Husserl's levels of constitution are based in his phenomenology of (human) experience, we would expect that they would correspond mostly to Fraser's higher levels of noo- and sociotemporality. This shows an important overlap between the two theories, in that both note the importance of individual and social "intentionality" (although Fraser and Husserl use this term differently[11]) to specific levels of time. Given Husserl's concentration on the relation of time to (human) consciousness, he provides us with multiple levels *within* Fraser's noo- and sociotemporal levels, offering nuances not dwelt upon in Fraser's general descriptions. On the other hand, Fraser's scientific background—and his desire to incorporate physics and the other sciences into a more universal understanding of time—results in his "lower" levels of nested temporalities, rarely touched upon by Husserl.

What I find most interesting, however, is the overlap of the hyletic flow with both the biotemporal and nootemporal levels. While Husserl's work separated the physical world of material objects from both the sensory world and the world of ideas,[12] his work on embodiment challenged this distinction, noting that the body is, in a certain way, a "turning point" between the physical and the sensory worlds.

---

was also a central interest, was Bergson (1859–1941), although including Bergson in our discussion would exceed the limits of this project.

11 Husserl's understanding of "intentionality" is very specifically related to consciousness: it is the directedness of consciousness toward something, the taking up of specific meanings or objects by consciousness. It is thus understood to be a "structure" of consciousness, since consciousness functions through being directed toward some type of meaning or content. For Fraser, this notion of "intentionality" is more diffuse, indicating a general direction toward an outcome (as in the "intentionality" of life at the biotemporal level) or a purposefulness (as in human apprehension of symbolic meaning at the nootemporal and sociotemporal levels). Thus, while there is some overlap in how each understands "intentionality," the concepts are by no means the same.

12 For Husserl, these are each separate "regions of being," discussed in a bit more detail in footnote 6.

> On the other hand, the Body appears here at the same time as a "turning point" where the causal relations are transformed into conditional relations between the external world and the Bodily-psychic subject. And in virtue of that, the Body appears as pertaining integrally to this subject and its properties, both the specifically Corporeal and the psychic ones bound up with them.
>
> HUSSERL 1989, 169

My body, in other words, is *both* a physical thing like any other, subject to temperature, sharp objects, gravity, etc., *and* it is a sensing being significantly different from any other physical object, simply because it is the only thing through which I am sensing. Thus, Husserl's (lowest) level of the hyletic flow becomes an important area for analysis: Not only does the hyletic flow implicate my body as a sensing being, but it is also the crux at which the physical world and the sensing body intersect—my body encounters the *physical* world *through sensing*. It is for these reasons that I find that the hyletic flow can be understood at both the nootemporal level (since sensing is perceived by us as thinking beings) and biotemporal level (since sensing is a connection to the living and physical world).

For Fraser, nested temporalities evolve or emerge from one another, the "lower" ones "subsumed" by the ones that follow: "Earlier temporalities are not replaced but subsumed by latter ones" (Fraser 2007, 48). In addition, he finds that the different orders of complexity and the different types of causation of each level confirms their distinctness: "The strikingly different orders of magnitudes of these numerical values [to measure complexity] strengthen the validity of the assumption of distinctness among them and supports the corollary assumption of level-specific Umwelten" (42). But if the hyletic flow bridges the gap between biotemporality and nootemporality, then these temporalities could be mutually influencing rather than—or perhaps in addition to—each being subsumed by the next.

For Husserl, the phenomenological levels filter into one another *in both directions* rather than one merely absorbing or building upon the other. As explained above, not only does sensory experience ground and feed into how we think and feel (as physical pain might affect our emotions or decisions about more abstract concepts), but conversely, how we think and feel emotionally can affect how we sense (as emotional engagement might distract us from, or heighten, physical discomfort). In addition, these mutual influences can go beyond my own sensations and thoughts, as intersubjective meanings can structure how individual and sensory input is taken up. Certain sensations can be highlighted or dismissed based upon cultural values and recognition, demonstrating how "higher" levels of experience or evolution may feed into the "lower" ones. While the latter point is primary in many contemporary theories in the humanities, for phenomenology I find that it is important to recognize *both* directions of influence.

For Fraser, layers *below* a specific temporal ontology will be implied in (or subsumed by) the current one, but higher layers remain out of the picture. Simply put, for Fraser, the higher levels do not play into the levels below them, even if the lower levels provide the necessary conditions for those that come above them. While this may make sense in evolutionary terms (the lower levels would exist before the higher ones, so the higher ones could not affect their development), the view is problematic for two reasons: First, Fraser's nested temporalities are not themselves *in time*. Rather, they are time itself, time's own evolution, a notion which Fraser acknowledges to be contradictory even as he justifies it (see Fraser 2007, 48 and 155), but which remains a sticky point in his theories. If the nested temporalities are not themselves in time, then there is no reason to assert that there must be a direction to their evolution, especially in how they influence one another. Second, even if we do allow that each temporality must emerge or evolve from those that are (ontologically) prior to it, once they have all emerged, there is no reason that they might not all be mutually influencing. One need only *imagine* an attacking tiger, or an angry boss, in order to experience increased heart rate, sweaty palms, headache, etc., which shows a nootemporal influence on the biotemporal level. Looking at sociotemporality, we find that many meanings developed at the societal level—as argued extensively in discourse, gender, race, and other critical theories—can affect, limit, and/or determine what I think, and even what I feel, at the noo- and biotemporal levels. The accumulation of life (and technology) on earth, moreover, is affecting the temperatures and even tectonic shifts of this planet—an effect of socio- and biotemporality on eotemporality—and speeding up processes that otherwise would have been much slower. This is not at all to say that the distinctions between temporal levels are dissolved. Rather, we find that the "higher" spheres can filter back into the "lower" ones in addition to the "lower" levels being subsumed by the "higher" ones. Limiting conditions, in other words, can go in *both* directions, not just the direction of temporal emergence. In this way, then, our comparison between Fraser and Husserl opens up a critical insight, suggesting that the temporal levels, however we conceive them, may overlap in more ways than originally described by Fraser, and that the influence from one level to another might move "downward" in addition to "upward."[13]

13 There is a limitation to this insight, implied in Fraser's descriptions of his "higher" levels (socio-, noo-, and biotemporality) as based in "intentionality," and his lower levels

The different levels of the hyletic flow, then, suggest that, while remaining distinct, temporal levels might influence one another in either direction. However, the idea that the hyletic flow overlaps the bio- and nootemporal spheres introduces a new consideration as well: while it is clear that the hyletic flow has a dual function, residing between the physical world and the sensing subject, it is not clear how well it really fits into the nootemporal sphere. The hyletic flow, in other words, while being the grounds for thought, is not really itself symbolic thinking. Thus we turn to the next analysis, which will look more carefully at the distinctions between the temporal levels, and at the biotemporal and nootemporal spheres specifically.

5 Critical Analysis B: the Function of Association

I would like to begin this analysis by focusing more rigorously on how Fraser describes the major shifts from one nested temporality to the next, especially from biotemporality to nootemporality. These evolutions are defined by distinctive transitions in causal relations (from chaos, to probability, to deterministic relations, to biological relations, to linguistic ability), as well as a major refocusing of material contents (light vs particles vs massive objects vs living matter vs symbolic thought). In other words, for Fraser, the move from one temporal level to the next constitutes a significant and paradigmatic change in how—and what—elements relate to one another, which thus brings about developments in time itself. In fact, he argues, while there must have been intermediary stages at some point, they no longer exist:

> There are, or at one time had to be, structures intermediate between those that comprise the known stable integrative levels. But such intermediate structures and functions either do not exist today or else are very rare. […] We know of no form of energy intermediate between the atemporal (light) and the prototemporal (particle). Also, matter is either particulate (prototemporal) or massive (eotemporal) by preference as it were, for almost all matter of the universe is in radiation, particles or galaxies. Going from the eotemporal to the biotemporal we note that biogenesis is

(eo- and prototemporality) in "causality" (deterministic and probabilistic, respectively). Husserl, too, relegated "causality" to the physical realm, but found that "motivations" have influence in sensory, conceptual, and behavioral experience. So, for *both* Fraser and Husserl, the causal world functions according to different rules than the world of thought and ideas. Nevertheless, we can still find examples that show mutual influence moving from one direction or the other.

> not now taking place spontaneously. Likewise, going from the bio- to the nootemporal umwelts, we have no living species intermediate between ape and man.
>
> FRASER 2007, 158–59[14]

So each temporal level stands on its own, with its own causal relations and types of objects, ultimately giving rise to the next level through a major transition in causation as well as a concentration and realignment of material. Each of these levels, further, has a "bottom" that is most closely connected to the level from which it emerged, and a "top" that is closest to the level that evolves out of it.

> Thus [...] the structural stability of our crystalline ancestors (on the top of the eotemporal world) survived in the stability, though not in the chemistry of the DNA molecule, on the bottom of the biotemporal world. Likewise, the behavioral pattern of the species that was between ape and man survived in a class of behavioral functions on the bottom of the nootemporal world. In depth psychology these functions make up the unconscious of man.
>
> FRASER 2007, 159

In this way, each of the temporal worlds has a range within it, one that reflects its evolution into the following level while also remaining within its own causal and material set.

I wonder, however, if the biotemporal level contains too broad a spectrum of types of existence. On its "bottom," as we see, lies the DNA molecule, as well as, perhaps basic cells and cell structures. At its "top" are animals with very high cognitive skills that lack extensive abstract language (for Fraser, the "apes"). Not only is this a huge range to fall under "biogenesis" and "living matter," but it is also a category that relies upon, what I see to be, two relatively unfounded assumptions: First, the "biotemporal" realm simply mirrors the academic division of "biology," which includes everything from studies of DNA to behavioral analyses of complex animals and herds. While all these areas certainly fall under the category of "living things," there is not much else that connects these relatively disparate fields of study. The DNA specialist usually cannot provide insight for the biologist in the field who is trying to figure out why elk are behaving in a specific way. Nor can the elk biologist often help

14 This once again brings out the tension in Fraser's descriptions of the temporal levels as both evolving into the development of time and as somehow in time themselves.

when the DNA specialist hits a snag in her research. This is not to say that these two researchers cannot communicate, nor that they could not conceivably carry out research together. Rather, it simply suggests that the "bottom" and the "top" of the biotemporal realm might not ultimately fit into the same sphere. The biotemporal level might be based upon the association of various areas under the term "life," but there may be no reason to regard this as a *single* ontological category. We need to look more carefully at the causes and materials at the bottom and top of the biotemporal sphere in order to determine whether they belong together. The second reason that I think the biological grouping might be suspect, is because it contains an implicit presumption that all pre-linguistic animals act through pure instinct, and that, further, instinct is biologically determined. For humans, according to Fraser, these instincts are buried in our unconscious, to be discovered through psychological analyses. But for animals, since they exist only at the biotemporal level, it is implied that they function solely through biological processes. Consider some of Fraser's descriptions of the biotemporal level:

> The level-specific causation of the biological Umwelt is short-term intentionality in the service of organic needs. For those who must breathe, it is a demand for air, for all organisms, the need for nourishment.
>
> FRASER 2007, 43

> The characteristic connectivity among events of the biotemporal world [...] is organic intentionality directed toward short term, concrete goals, serving the continuity of the life of an organism.
>
> FRASER 2007, 46

> The single-path, deterministic causation of the eotemporal world gave rise to multiple causation characteristic of life, as well as to final causation (goal-seeking behavior).
>
> FRASER 2007, 156

> Biotemporal events are joined by multiple and by final causation. Rigid programming gives way to dynamic programming.
>
> FRASER 2007, 157

While Fraser allows that animals have "dynamic" processes and "goal-seeking behavior," these actions and processes are still within the realm of causation and "programming"; it is only within the nootemporal level that we find freedom and creativity. "The biotemporal 'now' opens up to the mental present of man with its continuously varying boundaries, while the plurality of biological causations expands to include human freedom" (2007, 157). Implicit in these statements is another presumption: that all thought is symbolic, linguistic thought, which allows for conceptual thinking about ideas such as freedom. Accordingly, any "thought" that is not symbolic must be determined, not free, and is essentially *not* thought. In fact, Fraser allows that, while animals might sacrifice themselves for their young, it should be understood as instinct or programming: "Mother rabbits often protect their kits even at the price of their lives. They have great genetic investment in them and they are programmed to protect that investment" (2007, 169). My goal here is not to deny that animals (or even humans) often behave according to instinct (without thinking), nor to deny that human symbolic thinking is categorically different from instinct or genetic programming. I grant both of these claims. However, I would argue that many animals also function beyond mere deterministic goals laid out by genetics or instinct, and that they conceive and engage in the world in ways that go beyond mere survival mechanisms or biological processes—even if these ways are not yet fully symbolic thought. Many animals (including humans), in other words, engage in the world in a way that includes non-symbolic thought.

Without taking a deep dive into the various studies that have shown animal decision-making, planning, strategizing, group dynamics, and even language and emotions—well outside of the parameters of this project (and my own expertise)—I would propose that there might be a temporal level that reflects types of thinking not yet based in symbolic or linguistic meaning. Such a level might resemble Husserl's discussions of passive synthesis—and especially his concept of association—which point to a dynamic engagement with the world that is irreducible to biological processes. Further, association is a specific type of temporal relation that exceeds immediate biological goals; while certainly contributing to the survival of an animal or species, association also contributes to animal decisions and emotional connections, among other things.

As discussed earlier in relation to Husserl's level of "passive synthesis," the function of association indicates a fundamental type of relationship between the present and the past. For Husserl, association makes representational recollection possible, thus enabling the emergence of nootemporality, where we have an explicit relation with the past, present, and future. It is also an important structure of consciousness in general. Association relies upon the basic structure of inner time-consciousness in order to relate the contents of our experiences to one another. I am able to recognize similar or related objects, events, even contexts, because they have been retained, and because association links my present situation to past situations like it—and this recognition usually does not require explicit recollection. Association thus makes possible our habitual behavior, recognition, repeated skills, and much more. But this ability of association is not reducible to instinct, even if it might have originally developed for survival. It enables a pre-symbolic connection between events, abilities, and objects that connects us with our past, but which is not necessarily related to specific goal-seeking behavior or survival. Many types of animals apply past learning to current situations, create habits, recognize familiar objects, develop skills, foster complex relationships, etc. They respond to the meanings they associate with certain sounds, movements, colors, odors, and more. All of this shows a relationship with the past that is not linked to specific life processes, whether biological or instinctual. And while this may not precipitate conscious recollection in animals in the way that it can in humans, it demonstrates a mode of relationship between the past and the present that exists for both humans and animals, and—more importantly—indicates an important shift beyond the immediate biological temporality that sustains the continuation of life.

Given Fraser's arguments addressing how life and thought cannot be reduced to the eotemporal level of physical determinism, I would like to think that, in a parallel fashion and upon reflection, he would not simply relegate animals to pure biological processes, especially animals with complex skills, communication, and even social hierarchies. Nevertheless, I don't think that he would argue—and I wouldn't argue this either—that animals possess a clear sense of past, present, and future, nor that they engage with their temporal existence through the symbolic means found at the nootemporal level. Perhaps, then, we might add one more nested temporal level to Fraser's hierarchy, situating it between the bio- and nootemporal realms: We might call it the associo-temporal level. This level bridges the gap between the biological processes that define biotemporality and the abstract thought indicative of nootemporality. It provides a space for temporal engagement with our environment that is more than biological, but not yet conceptual thinking. Associo-temporal processes, which connect past experiences to present ones and serve as the basis for recognition, skills, and decisions, provide a depth to our relationship with the past beyond what is necessary for biological functions. And, like Fraser's other nested temporalities, it contains the grounds necessary for the temporal level above it, since association enables the link between past and present that makes possible a conceptual understanding of time as past, present, future.

But given this proposal, we must now examine whether the suggested associo-temporal level would *emerge* from the biotemporal one, and whether we can see how the nootemporal level would arise from associo-temporality. Biotemporality, as we have seen, introduces the now and direction in temporality, in a way that was unnecessary for the prior temporal levels. Fraser refers to this as the "short arrow" of time (Fraser 2007, 46). But association exceeds the immediate processes and short-term goals paradigmatic of biotemporality: it maintains a more distant, and often generalized past in connection with the present, and without the orientation of a specific future goal. Association provides a past example, or context, or relation that can be used to comprehend the present and decide how to move into the future. While this certainly does help with survival, it also extends into our discernment of likes and dislikes, emotional relationships, and behavioral habits unrelated to survival. Thus, association emerges from the short-term dynamics of continuing life, as seen at the biotemporal level, to a new level that maintains a broader sense of the past in relationship with the present and an open future. Causality is no longer biologically determined; rather, past events are given in relation to the present, motivating choices for the future. Just as, for Fraser, the biotemporal realm is "intentional," the associo-temporal level moves further into this intentionality by providing context. Here, animals (and humans) do not just react to their environments; they engage with them. It is here that the notion of "Umwelt" truly begins to take form. In fact, we might assign to the associo-temporal level a "medium-length arrow" of time, as compared with the "short arrow" of biotemporality and the "long arrow" of nootemporality.

On the basis of associo-temporality, then, we can see how nootemporality could arise. Symbolic thinking does not come solely from biological processes, but rather from contexts that call for abstract and conceptual thinking. These contexts are enabled through associo-temporality. In this way, nootemporality emerges from associative processes as a new intentionality that represents the past symbolically in order to create meanings and decisions about the future. Biotemporality, then, gives rise to (and is subsumed by) associo-temporality in that our biological processes and instincts for survival set up the parameters for a contextual awareness and relationship with a past beyond that given by specific biological functions. Then nootemporality arises from associo-temporality in that the associations and habits from the associo-temporal realm set up the conditions for symbolic and representative temporal relations. Each sphere has its own "causal" relations (biological/instinctual processes, vs. contextualizing associations providing motivations for future decisions, vs. symbolic thought following logical axioms), and each has its own set of objects (living matter, vs habits and relational contexts, vs abstract and conceptual thinking). In this way, associo-temporality both relieves biotemporality of the burden of carrying too much within it while also filling in some of the gaps between biotemporality and nootemporaliy.

In another sense, associo-temporality also addresses a concern addressed in the critical analysis above, regarding the relation of the hyletic flow to nootemporality. While the hyletic flow clearly grounds our conceptual thinking, it is much more closely related to our passively synthesized associations. Thus, it makes sense that the hyletic flow would still bridge two temporal spheres, but that these would be the bio- and associo-temporal spheres, rather than the bio- and the nootemporal. We might demonstrate this added level in our comparison between Fraser and Husserl in the following way:

Fraser	**Husserl**
Atemporal	–
Prototemporal	–
Eotemporal	–
Biotemporal	Hyletic flow (presence, duration, and succession)
Associo-temporal	Hyletic flow (presence, duration, and succession)
Associo-temporal	Passive synthesis (subjective relations of present, past, future)
Nootemporal	Active constitution (subjective time as object)
Sociotemporal	Interpersonal intersubjectivity (objective time)
Sociotemporal	Intersubjective community (abstract objective time)

The associo-temporal level, as we can see, relieves the biotemporal level from having to maintain basic life processes as well as the complex relations of developed animals with their environment. Instead, the associo-temporal level covers the hyletic flow and the passive synthesis of association. These already have a close relation in Husserl's analyses as both address types of association, and function primarily without our focused awareness. The nootemporal level, furthermore, is now more accurately connected with active constitution, both of which describe the symbolic and meaningful relations that humans have with the world, usually through language. In this way, then, bringing Husserl and Fraser together provides new insights, and suggests possible modifications, which can help us with our overall understanding of time.

6 Conclusion

To conclude, a comparison between Fraser and Husserl does not just give us an outline of how their "hierarchies" might overlap. Rather, it provides openings to further insights, especially the mutual influence of each of these levels on the others, and the possible introduction of an additional level between Fraser's bio- and nootemporality, called associo-temporality. The latter would accommodate the temporal experience of animals as well as our own non-cognitive temporal experiences. The former, the mutual influence of the levels on one another, might also be a way to address criticisms that see Fraser's nested hierarchy as both rendered in time and as time itself. Nevertheless, these analyses were meant only to introduce a possible discussion between Fraser's and Husserl's works—a dialogue that appears promising based upon what we have accomplished here, and one that might continue to provide critical insights to further our overall project of understanding time itself.

References

Fraser, J. T. 2007. *Time and Time Again: Reports from a Boundary of the Universe.* Leiden: Brill Academic Publishers.

Husserl, Edmund. 1983. *Ideas Pertaining to a Pure Phenomenology and to a Phenomenological Philosophy, First Book: General Introduction to a Pure Phenomenology* (*Ideas I*). Translated by F. Kersten. The Hague: Martinus Nijhoff Publishers.

Husserl, Edmund. 1989. *Ideas Pertaining to a Pure Phenomenology and to a Phenomenological Philosophy, Second Book* (*Ideas II*). Translated by R. Rojcewicz and A. Schuwer. Dordrecht: Kluwer Academic Publishers.

Husserl, Edmund. 1999. *Cartesian Meditations: An Introduction to Phenomenology.* Translated by Dorion Cairns, Dordrecht: Kluwer Academic Publishers.

Husserl, Edmund. 2001. *Analyses Concerning Passive and Active Synthesis: Lectures on Transcendental Logic.* Translated by Anthony Steinbock, Dordrecht: Kluwer Academic Publishers.

Rodemeyer, Lanei. 2021. "Levels of Embodiment: A Husserlian Analysis of Gender and the Development of Eating Disorders." *Time and Body. Phenomenological and Psychopathological Approaches*, edited by Christian Tewes and Giovanni Stanghellini, 234–255. Cambridge University Press.

CHAPTER 3

Leibniz and the Genius of Time

Walter Schweidler

Abstract

Leibniz was the philosopher (and the inventor of the infinitesimal calculus) who, as Ernst Kantorowicz said, "turned the *aevum* from heaven to earth." In his concept of the "monad" Leibniz examined the measurability of time through the relation between the infinite continuity of steadily passing time and the finite totality of all the identifiable beings in the real world. The monad is the factor that divides the determinate order of reality from the unmeasurable "ocean of possibilities." Like the medieval "angel" and its antique predecessor, the "genius," the monad thus addresses the crucial paradox that time precedes as well as presupposes its measure. One impressive architectonic illustration of this appears in Leibniz' lifelong dedication to the planting and design of the famous landscape park in Hannover-Herrenhausen. It was Leibniz' idea to make this garden a monument of typically baroque symmetry combined with one exceptional factor, which he characterized as the "art of deviation." It is this factor, invisible to the spectator, but determining the whole, which allows, as Leibniz wrote to Clarke, the existence of temporal development in the world.

Keywords

Leibniz – Herrenhausen – monad – genius – angel – infinity – totality – imperceptibles

"Each portion of matter may be conceived as like a garden full of plants or like a pond full of fishes. But each branch of every plant, each member of any animal, each drop of its liquid parts is also some such garden or pond" (Leibniz 1998, 256).[1] This somewhat cryptic passage from his *Monadology* (§ 67) is rooted in Leibniz' onto-theological metaphysics and in the conception of the universe which laid the groundwork for his invention of infinitesimal calculus. The

1 See also Leibniz (1991, 213) and Leibniz (1956, 59).

 | DOI:10.1163/9789004723634_005

philosophical task which the scientist and mathematician Leibniz sought to accomplish throughout his lifelong conceptual struggle was to demonstrate the compatibility of totality and infinity. His general idea was that the universe must be seen as composed of components which are infinitely divisible, but which also form a complete whole according to irrevocably determined laws of nature. Infinity is required in order to explain why the infinitesimal calculus can connect all parts of our world; totality is demanded as the decisive criterion that distinguishes this one and only real world from the "ocean" of all other possible worlds, worlds which do not exist because their components lack the optimal and complete consistency that could form them into closed whole. According to this general idea, reality is constituted by a perfect and unreproducible harmony; another "world" than the one in which we live would be either not real or identical with this one. Harmony is thus the ontological principle of reality and of, what in the end is the same, truth.

This conception of harmony as the constitutive structure of the world does not depend on mere metaphysical speculation or religious belief. It is rather a principle quite consistent with common sense and our everyday experience. When we tell the truth, everything fits together; we can be sure that, as long as our memory does not fail us, there will be no contradiction in what we say. When, however, we lie, we soon find we have to invent, asking ourselves what must also have happened if what we pretend to have happened had occurred. Harmony in this way signifies the ultimate difference between falsehood and reality. And the reason for this is, according to Leibniz, constituted by infinity, that is by the "infinite ocean" of possibilities that arise as soon as we try to construct a world any different from the one in which we live—in which, for example, only one single event took place one second or one half second or a hundredth of a second earlier or later. It is impossible to ascribe totality to any such world, or to any world other than the one real world. We can ask what happened between Lady Diana's leaving the Ritz and her fatal accident, although we may never find the exact answer. But we cannot ask what happened between Scarlett O'Hara's escape from Atlanta and her return to Tara, not because we lack the capacity of finding the answer, but because the question is nonsensical. Facts only exist in this one world. If we try to imagine a world in which these facts are repeated, then the object of our imagination would be identical to this one real world. This is the content of Leibniz' fundamental principle of the *identitas indiscernibilium* ("identity of indiscernibles"), the principle that demands a mediation between totality and infinity. Since any philosophical account of infinity that explains the world by means of mathematics must include infinity within, and not beyond, the world's totality, this principle is also crucial to Leibniz' system.

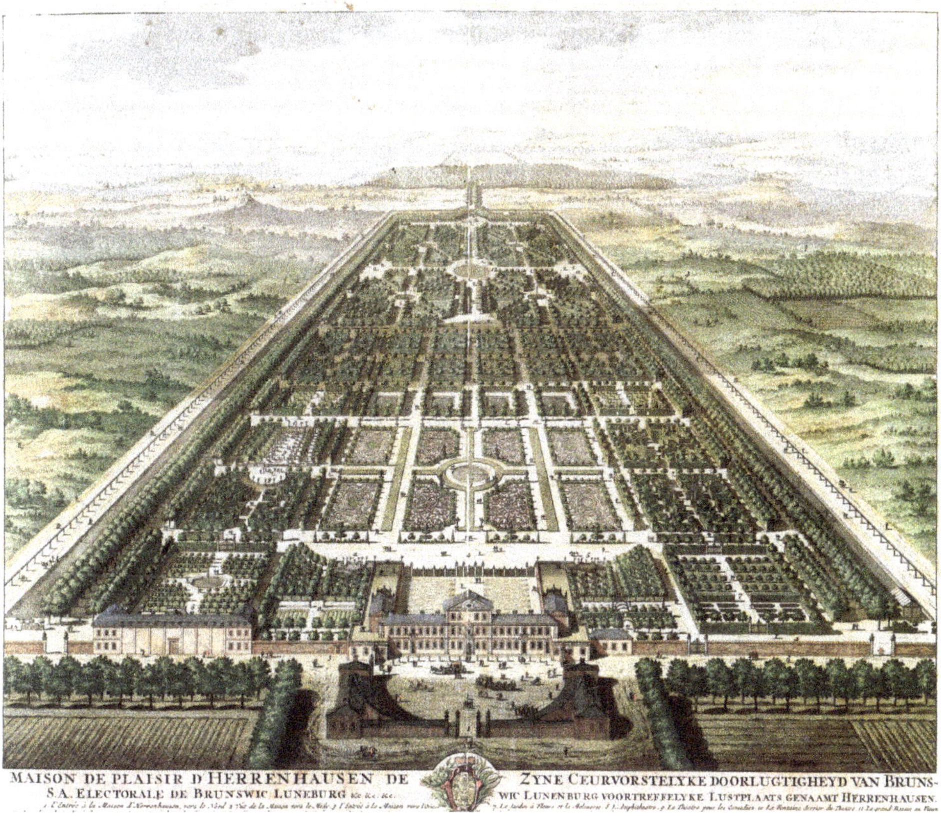

FIGURE 3.1 The Great Garden at Herrenhausen, ca. 1708
PAINTING. WIKIPEDIA, ACCESSED MARCH 27, 2025, HTTPS://EN.WIKIPEDIA.ORG/WIKI/HERRENHAUSEN_GARDENS#/MEDIA/FILE:UNBEKANNT,_MAISON_DE_PLAISIR_D'HERRENHAUSEN,_C1708..JPG

We must now turn to a circumstance that may seem at first glance to be merely anecdotal, but which is in fact crucial for an understanding of both the identity of indiscernibles and the "monad," a concept which lies at the center of Leibniz' metaphysical model. In his *Nouveaux essais sur l'entendement humain*,[2] a treatise which would become highly influential in the epistemological debates of the 18th century, Leibniz emphasizes that a decisive source for the discovery of the identity of indiscernibles were the observations he made over the more than three decades in which he was occupied with the planting and design of the landscape park in Hannover Herrenhausen.

2 For an English translation, see Leibniz (1996, ch. 27, par. 3).

This huge park (800 meters long and 450 meters wide), named the "Große Garten," is the core of an ensemble of three gardens belonging to the Herrenhausen castle, the summer residence of the Guelf dynasty, which had become one of the most important princely families in Germany in the second half of the 17th century. The construction of the Herrenhausen park began in the middle of the century and was essentially finished around 1700. The park's form, which still remains, was largely determined by the Duchess Sophie, granddaughter of the English King, James I, and wife of the Duke and "Kurfürst" Ernst August von Hannover.[3] The Duchess Sophie and her daughter Sophie Charlotte, the wife of the Prussian king Friedrich I, were also Leibniz' most important discussion partners after the Duke appointed him designer of the garden in 1696. The *Nouveaux essais* recall the conversations Leibniz had with the two ladies on their long walks in the park, conversations that provided the basis for the principle of the identity of indiscernibles. One scene especially, from the year 1705 (and depicted on an engraving 90 years later), was described by Leibniz in a 1716 letter to Samuel Clarke.[4]

The scene shows him in debate with Sophie, her daughter, and a nobleman, Carl August von Alvensleben.[5] Alvensleben had insisted that if one searched long enough one could find two leaves of completely identical shape in the garden. Leibniz disagreed, and, at least for the few hours that the search lasted, turned out to be right. They never found one leaf absolutely similar to another. It was this episode that Leibniz later recalled as leading him to conclude that there are no atoms, nor a vacuum, nor any "absolute" time beyond or separable from the events of the universe. He did not explicitly specify the connection, but the general idea must have been the principle of the identity of indiscernibles.

So, here in Herrenhausen, Leibniz had found a paradigm for the identity of indiscernibles, that is, for the absolute individuality of each and any life-form, including the organism of the world-whole, the universe. What we now know about fingerprints, for example, was what Leibniz explored in garden leaves: that there is an absolute and unrepeatable individuality and uniqueness in every specimen of living nature. Biologists such as Ernst Mayr now ground this principle on the insight that nature itself is the result of history and that all structure within it reflects a development composed, in Wittgenstein's words, not of things but of facts. Whatever belongs to our universe has its own unique

3 The Duchess' son, Georg Ludwig, ascended the English throne in 1714 as King George I.

4 See Bredekamp (2012, 142).

5 One of von Alvensleben's descendants has written one of the most important books on the current research being conducted on the Herrenhausen park, see Alvensleben (2021).

FIGURE 3.2 Drawing. In *Leibniz und die Revolution der Gartenkunst*, by Horst Bredekamp, Herrenhausen, Berlin 2012, 34

place in time or, as we might put it, its proper time-place.[6] This general idea helps explain Leibniz' cryptic description of the branch which is also a garden and the liquid drop which is also a pond. And, in fact, the idea is key to Leibniz' ontological explication of the measurability of time. To advance one step further, however, we will have to bring in the passage which follows immediately, in § 68 of the "Monadology": "And though the earth and the air which are between the plants of the garden, or the water which is between the fish of the pond, be neither plant nor fish; yet they also contain plants and fishes, but mostly so minute as to be imperceptible to us" (Leibniz 1998, 256). The epistemological category of the "imperceptible" (for us) thus takes on an essential role in explaining what it might mean that a drop of liquid in a pond "is" (in itself) a pond.

Here we must turn to the concept of the "monad." Like Newton, Leibniz was convinced of the existence of natural laws which completely determine the course of nature. These laws make mathematics applicable to the real world and allow us to measure space and time through methods based on differential and integral calculus. Leibniz, however, came to the conclusion at a relatively early stage of his intellectual development—almost against his will, as he admitted in his 1686 "Discourse on Metaphysics,"[7]—that the real components of the universe, the parts united in the harmonic totality of all time-places, cannot be conceptually reconstructed by means of our mathematical and scientific methods. In Aristotelian terms, motion, *kínesis*, is a continuous process, but *genesis* and *phthorá*, generation and destruction, are manifestations of discontinuity. The change from a living body to a corpse follows the same natural laws as any other change, but the person who has died has not changed his or her constitution, but rather no longer exists in the natural world. When two persons conceive a third one, the course of nature is not simply continued; rather a new being has been added to the world-whole as an additional, unique factor within it. What exactly defines the uniqueness of such an additional factor is not a scientific, but a philosophical problem, and Leibniz' solution is concentrated in the concept of the "monad."

Monads are the last units of the world-whole, and it is important to remember that the "whole" here must be understood in a genuinely spatio-temporal sense. The totality of the universe is composed by the process, we can even say, by the history of its composition, and nature is as much the result of this process as it is its principle. Time itself, including its measurability, is rooted

6 I have developed the concept of "time-place" in *Wiedergeburt*, see Schweidler (2020, 396 and 2022, 7). A short explication can also be found in Schweidler (2023, 56).

7 See Leibniz (1988, §§ 9 and 11).

in this natural process. As a result (so Leibniz in the preface to the *Nouveaux essais*) the present is "pregnant" with the coming events, and the future can be "read" in what is going on now. So, what we perceive when we are confronted with natural objects is also the history through which they have come to be. When we look at the stars in the night sky, although we perceive only the light that emanated from them millions of light-years ago, we nevertheless "see" the stars, although we are looking into the past. In this sense, the light contains the stars from which it came. And in this sense as well, the branch in front of us also documents the chain of being connecting back to the tree from which the seedling was taken, that then grew up to be the present tree that has produced this branch. Therefore we can, in a certain way, say that the branch contains the tree and that the fish contains the pond of which it is part.

Here epistemology and ontology coincide. What I perceive now contains the past that has made it perceptible to me, but this is so because I *am* as one unique factor in the world, a "monad" related to all other monads within the one unique universe. Leibniz hereby reactivates the Aristotelian dictum that the soul is "in some way everything," *anima quodammodo omnia*. And he reactualizes the Aristotelian principle that our self-experience must be considered the paradigm for any relation between the factors of the world-whole. Leibniz expresses this in saying that "every Monad is, in its own way, a mirror of the universe" (Leibniz 1991, § 63). If we could follow any monad's relations to the rest of the world far enough, we would in the end gain a perspective on the whole universe. Here we also see that infinity and totality are mutually dependent: "the Author of nature has been able to employ this divine and infinitely wonderful power of art, because each portion of matter is not only infinitely divisible, as the ancients observed, but is also actually subdivided without end, each part into further parts, of which each has some motion of its own; otherwise it would be impossible for each portion of matter to express the whole universe" (Leibniz 1991, § 65).

Epistemology and ontology thus coincide in the definition of the monad and its paradigmatic role for the understanding of the world-whole: infinite divisibility is an ontological category, while the idea that the universe consists of infinitely divisible parts has epistemological conditions. I can claim to be a universal mirror of the universe to which I belong because the relation that connects all the factors of the universe is also the relation which connects my body, as just one among all the parts of the world-whole, with my soul, as that which is "in some way everything," *quodammodo omnia*. This relation, which makes the harmony of the universe re-presentable in me, even though I am only one present being among all the others, is, according to Leibniz, *the harmony between the soul and the body of a living being*. As Leibniz puts it, "as the universe is ruled according to a perfect order, there must also be order in that which represents it, i.e. in the perceptions of the soul, and consequently there must be order in the body, through which the universe is represented in the soul" (Leibniz 1991, § 63). The insight that a monad "expresses" or "mirrors" the whole is based on the one fundamental factor that inserts into the whole of the universe the capability to *represent* it. Or, to put it differently, the ability to *re-present* allows the result of giving birth to the future to be already present in the present, so that the birth of the future is always already an act of rebirth.

As speculative as this may sound, it can at least shed light on the significant religious and mythological background of the concept of the "monad." As both the passages of the *Discourse on Metaphysics* cited above, and philosophical-historical research have shown,[8] the monad is in many ways the secular heir to the Medieval "angel" situated in the "aevum" as the temporal element between divine eternity and human finiteness. According to Ernst Kantorowicz in *The King's Two Bodies*, Leibniz "turned the *aevum* from heaven to earth" (1992, 291; my translation). The angel had been defined by Thomas Aquinas[9] as a created being who, unlike the rest of nature, is not composed of matter and form, but is rather a purely spiritual entity whose individuality is determined by its form alone, that is, an angel's identity is identical with its form. This implies that there can be no causal influence of one angel upon another—as in Leibniz' famous statement that "Monads have no windows" (Leibniz 1991, §7). As a result, the relation of one angel to another and to the rest of the universe must have been fixed by the creator from the beginning of creation and must remain unchangeable until its end.

In this way angels represent the fullness and the end of time within their created, and therefore temporal existence. Further, one particular kind of angel, the Guardian angel, who is born with and protects one individual human person throughout his or her life, views its ward through the perspective of the end of his or her days. The Guardian angel, however, is also the heir of a much older figure, the antique *genius*,[10] that we commemorate, according to the Romans, when we celebrate our birthdays. The Roman rhetor Censorinus, in his book *De die natali*, defined the genius as the god who is responsible for our birth (*ut geneamur curat*), is born with us (*una genitur nobiscum*), and receives and protects us (*nos genitos suscipit et tutamur*). The genius enables us to measure the time allotted at our birth and to relate it to the whole of the universe. When we celebrate our birthday, we do not express satisfaction that we have come one

8 See especially Schönberger (2003, 73–90).
9 See Aquinas (1981, q. 50, art. 4).
10 For a discussion of this concept, see Schilling (1978).

year closer to the end of our life. Rather we are honoring the progress we have made towards the full re-presention our life makes within the ensemble of all the unique factors that constitute the composition of the whole. One could say that by defining the final units of reality as "monads" Leibniz has—via the Medieval "angel"—ascribed to every part of the world-whole a kind of "genius" by which it anticipates the fulfillment and completion of the universe and to which it owes its generation and its presence among existing beings.

We can now try to point out the relation to Leibniz' Herrenhausen project. It is exactly the aspect that we saw in "Monadology," § 68 concerning the "imperceptible" but still real relation through which a part is able to contain the whole to which it belongs and from which it derives. As we have now seen, this can be explicated in an essentially temporal way. The *time* of the universe is that in respect of which one can say that the seedling contains in itself the tree from which it was taken. In this way "mirroring" is constitutive of the relation of being "contained" within time. What, however, is constitutive of the mirroring itself? Leibniz' answer must certainly refer to the essential laws which determine the course of the world. We "mirror" the universe by discovering and formulating these laws, especially through infinitesimal calculus. But the laws, and with them the relations between the parts of the world determined by the laws, were, are and will be there independently of our existence. So, what we consciously perceive, as well as how we describe our world on the basis of that perception, is only one side, the conscious side, of how we mirror the universe. The other side is the imperceptible one, and to avoid confusing the two, Leibniz called the conscious mirroring *apperception* (Leibniz 1991, § 14), while he ascribed *perception* to every monad, conscious or not. Perception, Leibniz says, "is to be distinguished from Apperception or Consciousness," emphasizing his anti-cartesian point by adding that "In this matter the Cartesian view is extremely defective, for it treats as nonexistent those perceptions of which we are not consciously aware" (Leibniz 1991, § 14). The distinction between the two sides, however, is decisive for what Leibniz calls the "harmony" between our body and soul. To come to a philosophical explication of this harmony, however, we must further ask how we can consciously represent that which is by definition unconscious? In order to become conscious of the relation between the conscious and the imperceptible sides of our "mirroring," we would have to achieve a paradox: we would have to make the imperceptible perceivable. And that is exactly what Leibniz wanted to achieve and to visualize with the symbolism of the "Große Garten." Leibniz needed such a gigantic facility to combine the two essential elements of his philosophical project, the perceptible and the imperceptible. The baroque garden was a genuine work of art that could represent the harmony of infinity and totality and also include

FIGURE 3.3 Drawing. In *Leibniz und die Revolution der Gartenkunst*, by Horst Bredekamp, Herrenhausen, Berlin 2012, 74

the one unique, exceptional and additional factor that allows for a perception of the imperceptible. Leibniz' strategy achieved both results. Firstly, the selection and design of plantings combines the strict symmetry of the formal elements (ponds, woods, parks, fountains, statues) with a deliberate diversity of individual plants. Every hedge corresponds to its symmetrical counterpart, but the plants from which it is composed are completely individual and different.

The whole ensemble is thus an illustration of the insight that infinity must be sought within, not beyond, the world-whole. This insight also corrects a political misinterpretation of the Garden's symbolism. The line of sight by which the visitor's view is led through the entire garden is not designed to represent an endless axis or, as in a common misconception, to symbolize the infinite power of the royal spectator. Even the "Sun-King's" park at Versailles, which was the ideological model for Herrenhausen, is presented in contemporary paintings with a very long, but clearly not unlimited axis. Like its French model, the Herrenhausen park does not try to create a *trompe l'oeil* in order to deceive the spectator. Rather the park represents the *coup d'oeil*, the

"Augenblick" in which visitors survey an intrinsic infinity which they, as finite beings, can perceive only in part, but which they nevertheless recognize as a closed whole. To borrow from the terminology of art history, Herrenhausen is a *hortus conclusus* ("enclosed garden"), a renewal of the Garden of Eden, which does not intend to place the visitor in the position of God, but rather to open his eyes to a divine perspective which may lead him to God.[11]

But let us come finally to the factor inserted into the scenery in order to make the imperceptible perceivable. This factor is time. According to Leibniz, time is determined by an exactly and seamlessly measurable structure, but the *origin* of temporality escapes all measurable order; it is a jump in nature, a disruption to its orderly course, even, as we would probably say nowadays, a mutation. The most original and surprising feature in Leibniz' Herrenhausen architecture demonstrates this. This is a feature that no spectator can consciously perceive, but which was obviously a directing and highly demanding (and expensive) guideline followed over the decades required by the design of the scenery: the garden is not, as it appears to be, rectangular.

Leibniz characterized the architectural strategy behind this—a strategy which he probably did not invent but found already at work when he began his activities—as the "art of deviation." He saw the (meta-)physical basis of this deviation in a factor which leads us even further back in the history of science than the idea of a personal "genius." Leibniz described the effect with a term used by Lucretius and Cicero, as *inclinationes*, imperceptible tendencies one is not aware of,[12] but by which one is mentally directed. His own term for these mental powers was "*petites perceptions*," the contents of our own thinking which form our knowledge without coming into our consciousness. The *Nouveaux essais* refer several times to these unperceived perceptions, for example, a machine making noise that we notice only when it ends, or bell chimes which, after we have turned our attention to them, we can still count even when they are over. Leibniz employed these phenomena in arguing against both John Locke and Descartes, seeing these *petites perceptions* as the connecting link between our conscious perceptions and the *appetitions* which direct our consciousness on an imperceptible level, defining "appetition" as the "activity of the internal principle which produces change or passage from one perception to another" (Leibniz 1991, 15).

The continuity with Medieval metaphysics is obvious. *Perceptio* and *appetitus* were exactly the two capacities of the angels which, according to Saint Thomas, constitute their relationship with each other and the rest of the

11 For this interpretation, see Bredekamp (2012, 75–81).

12 See also Leibniz (2012, book 2, chapter 2).

FIGURE 3.4 Ground plan of the actual park. Drawing. In *Leibniz und die Revolution der Gartenkunst*, Berlin 2012, 87

world. The only addition that Leibniz makes in the *Monadology* is the principle that holds them together: infinity. Our conscious perceptions and our appetites are, as he wrote in a letter in 1715,[13] composed of infinitely many inclinations of which we are unaware. And in his 1716 letter to Clarke[14] he states that this imperceptible factor, invisible to the spectator but determining the whole gigantic architecture of our thought, represented that without which there could be no temporal development at all.

This temporal factor thus explains Leibniz' cryptic remark in *Monadology* § 68. Our "mirroring" the universe has both a conscious and an imperceptible side that corresponds to an ontological aspect. Just as we measure time through an infinitesimal calculus, a dynamic development at work in the inclinations leads our perception on an imperceptible level and forms the connection between infinity and totality. Like the two sides of a leaf, the mathematically structured measure of time is inseparably connected with the whole living history of the universe, from its origin to the present moment. The "art of deviation" represented in the Herrenhausen park symbolizes a factor that escapes mathematical methods of measuring time, but that discloses the "object" of that measurement, which is *time itself*.[15] Maurice Merleau-Ponty's dictum that what "does not pass in time is the passing of time itself" (2005, 492) or Emmanuel Lévinas' characterization of time as "that astonishing divergence of the identical from itself," (1993, 98) is also the ontological key to Leibniz' principle of the identity of indiscernibles. Our world receives and sustains its real existence by its living and creative deviation from itself. By measuring time we experience the self-resemblance of our universe as it catches up with its own fulfillment.[16]

References

Aquinas, Thomas. 1981. *Summa Theologica*. Notre-Dame, Indiana: Ave Maria Press.

von Alvensleben, Udo. 2021. *Die Braunschweigischen Schlösser der Barockzeit und ihr Baumeister Hermann Korb*. Berlin: De Gruyter.

13 See Bredekamp (2012, 90 and 143).

14 See Bredekamp (2012, 142).

15 This conception can be traced back to Pindar's topos of the "turning of time" and the platonic "*exaiphnes*"; see also Schweidler (2023, 16 and 18).

16 For the relation of the self-resemblance of the world to the understanding of temporality see Schweidler (2022, 28, 490, 496 and 501).

Bredekamp, Horst. 2012. *Leibniz und die Revolution der Gartenkunst. Herrenhausen, Versailles und die Philosophie der Blätter*. Berlin: Wagenbach.

Kantorowicz, Ernst. 1992. *Die zwei Körper des Königs. Eine Studie zur politischen Theologie des Mittelalters*. Stuttgart: dtv.

Leibniz, Gottfried Wilhelm. 1956. *Prinzipien der Natur und der Gnade. Monadologie*. Hamburg: Meiner.

Leibniz, Gottfried Wilhelm. 1988. *Discourse on metaphysics and related writings*. Translated and edited by R. N. D. Martin and Stuart Brown. Manchester: Manchester University Press.

Leibniz, Gottfried Wilhelm. 1991. *The Monadology: An Edition for Students*. University of Pittsburgh Press.

Leibniz, Gottfried Wilhelm. 1996. *New Essays on Human Understanding*. Translated and edited by Peter Remnant and Jonathan Bennett. Cambridge: Cambridge University Press.

Leibniz, Gottfried Wilhelm. 1998. *The Monadology and Other Philosophical Writings*, edited by Robert Latta. Oxford: Clarendon Press.

Leibniz, Gottfried Wilhelm. 2012. *Nouveaux Essais*. Edited by Emile Boutroux. Neuilly sur Seine: Ulan Press.

Lévinas, Emmanuel. 1993. *Otherwise than Being, or Beyond Essence*. Translated by Alphonso Lingis. Berlin: Springer.

Merleau-Ponty, Maurice. 2005. *Phenomenology of Perception*. Translated by Donald A. Landes. London: Routledge.

Schilling, Robert. 1978. "Genius," *Reallexikon für Antike und Christentum*, volume 10, edited by Theodor Klauser. Stuttgart: Anton Kiersemann Verlag.

Schönberger, Rolf. 2003. "*Monas, id est unitas*. Thomas' Angelologie und Leibniz' Monadologie." *Monadisches Denken in Geschichte und Gegenwart*. Edited by Sigmund Bonk. Würzburg: Köningshausen & Neumann.

Schweidler, Walter. 2020. *Wiedergeburt*, Volume 1. Freiburg/München: Alber.

Schweidler, Walter. 2022. *Wiedergeburt*, Volume 2. Baden-Baden: Alber.

Schweidler, Walter. 2023. *The Other Time. Philosophical Approaches to the Past that has Never been Present*. Baden-Baden: Alber.

CHAPTER 4

Measuring and Taking the Measure of a Stone's Time

Paul A. Harris

Abstract

This essay presents a transdisciplinary attempt to measure a stone's time. The endeavor entails more than measuring the rock's age—dating the origin of mineral material using scientific methods yields a fact, a (very large) number of years. But to know the geologic age of a rock is not the same thing as understanding its 'time' in any intuitive or visceral—let alone philosophical—sense. Perhaps it is more fitting to speak of 'taking the measure of a stone's time': to take the measure of something is to delve into and discover its nature, to develop a deeper insight into or sympathetic or empathetic feeling for it. I will take the measure of a stone's time by taking different passes at and with the stone, integrating methods and materials from philosophy, geology, and artistic practices. This essay invites the reader to suspend judgment, to participate in speculative thinking or a kind of thought experiment.

Keywords

deep time – slow time – geology – amphibolite – Suiseki – timefulness – Gaston Bachelard – material imagination – Marcia Bjornerud – Roger Caillois – silence

This stone is an amphibolite. It exudes a calm quality; it is dark, dense, and very smooth. It is a large river rock (measuring 48 cm/19″ L × 13 cm/5″ W × 8 cm/3″ D) and surprisingly heavy for its size—when I bring this stone to presentations, people passing it around often almost drop it because of its unexpected weight. I found this stone in 2019 in the Cache de la Poudre River in the Rocky Mountains of Colorado, a designated 'wild and scenic' river that has carved a narrow, winding canyon that drops 2,100 m/7,000 m in elevation and features striking rock formations. As architect and landscape artist Charles Jencks has observed, searching rivers for interesting stones is "an enjoyable pastime in itself," one which "deserves celebration and new appellation, 'rock fishing'"

 | DOI:10.1163/9789004723634_006

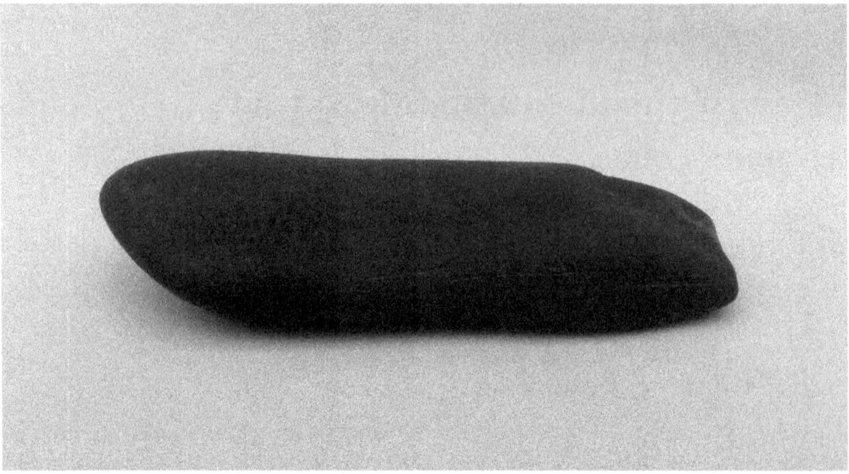

FIGURE 4.1 Amphibolite
PHOTO BY PAUL HARRIS

(2005, 190). On this particularly beautiful autumn day, I had one of those rare experiences that rock fishers recognize and often speak of: walking along and suddenly feeling a kind of signal to one's attention or pull in one's body, and finding that it came from a specific stone. On this occasion, I turned and stepped into the cold water, leaned down, and pulled the amphibolite out of the river. In the ensuing years, I have developed a deep attachment to this stone. I kept this stone at hand on a table when I lived in Colorado and I continue to keep it next to my desk in my study in California. I have spent hours looking at, touching, holding, and thinking about this stone. My experience of and relationship with the stone are expressed beautifully by Carl Austin Hyatt in his essay "Listening to Stone Beings":

> Over the years, I found myself called to individual stones they seem to exude a palpable numinous quality they felt like heavy, textured pools of silence holding a visceral stillness that resonated with some silent awareness within me. There was a sense of shared recognition which deepened my confidence in our connection.
>
> HYATT 2021

As a professional photographer and artist, Hyatt attempts to understand stones by photographing them in his studio. He attests that "their felt yet unseen presence started to become the real subject of my photographs" (2021). I also find photographing stones a productive means of seeing stones in new ways, particularly in tandem with practices of stone display. The amphibolite in particular lends itself to display as Suiseki, a Japanese tradition I have learned about through my involvement with California Aiseki Kai, a Suiseki and viewing stone club founded in 1983 by Larry and Nina Ragle in Pasadena, California. The viewing stone club was an offshoot of a Bonsai group founded by John Naka, a renowned Bonsai cultivator and teacher. As described on the California Aiseki Kai website, *Suiseki* "are small stones shaped by nature which suggest familiar landscapes such as mountains, islands, waterfalls, shorelines or seascapes. To be classified as *suiseki*, the material must be dense, dark in color and possess a subtle patina" (Aisekikai). Suiseki are traditionally displayed in hand carved wooden bases, called *daiza*, or in trays of sand, called *suiban*. The resemblance of the amphibolite from Colorado with river stones in Japan makes sense, as both environments were formed by volcanic chains. My first formal display experiment was to set the amphibolite in a *suiban* I had on hand to bring out its subtly curved form, suggestive of a landscape.

From an aesthetic standpoint, this display does not do justice to the stone. The *suiban* is too small and the stone sits up on the sand. Rather than creating the scalar effect of a stone projecting a landscape from a tray, the display looks like a rock resting on some sand. In the hopes of creating a display worthy of being shown in California Aiseki Kai's annual viewing stone exhibition at the Huntington Library and Gardens, I took the stone to the home of Jim Greaves, Curator of the American Viewing Stone Resource Center at the Huntington Library. Different displays of a stone bring out different affects, changing the energy or presence of the stone. Setting the amphibolite vertically, for instance, transformed the serene stone into an assertive statement of lithic power.

For the Huntington exhibition, we settled on a traditional display of the stone in a *doban*, a bronze tray, that provided a stronger ground for it. I gave it the title *Touchstone of Silence*.

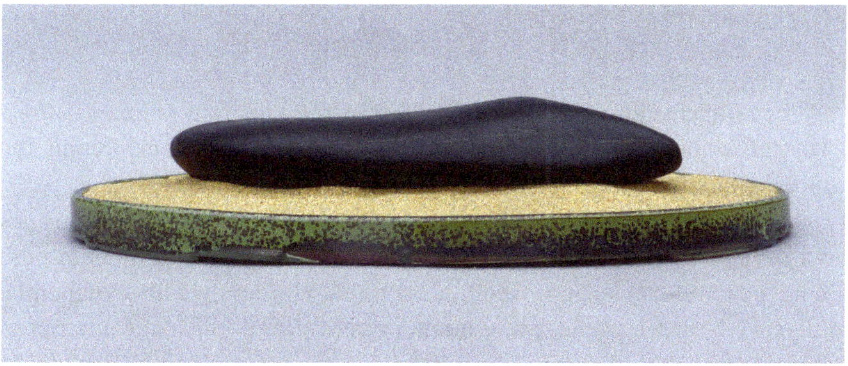

FIGURE 4.2 Amphibolite in *suiban*
PHOTO BY PAUL HARRIS

FIGURE 4.3
Amphibolite vertical display
PHOTO BY PAUL HARRIS

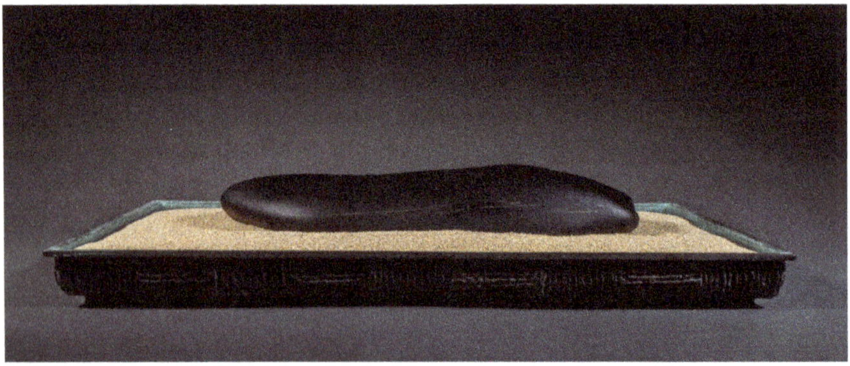

FIGURE 4.4 Touchstone of Silence, Huntington Library and Gardens Viewing Stone Exhibition, December 27–31, 2022
PHOTO BY ERIC STONER

When Touchstone of Silence appeared in the Japanese magazine *Aiseki* ('Love of Stones') in June 2023 in a feature on the Huntington exhibition, the amphibolite gained international renown.

Having explored and brought out the aesthetic and affective dimensions of a stone, how does one measure its time? In temporal terms, stones pose an enigma and challenge to the human mind. The irreducible, exponential difference between geologic and human timescales ruptures any attempt to forge relations with the stone, disrupting even the most speculative modes of thought. The more one considers geologic time, the more it seems to recede. As Jeffrey Jerome Cohen writes in *Stone: An Inhuman Ecology*, "Lithic-induced perspective shift triggers an ontological and temporal reeling, a rocky movement of affect, cognition, horizon" (2015, 16). Nonetheless, the place to begin in taking the measure of a stone's time is through a geologic account of its composition and history. The geologic timescale, calibrated collaboratively over more than a century, is a marvelous achievement that provides a comprehensive history of the Earth, divided into eons, eras, and periods measured in millions of years. A study of Earth history and its cycles and tempos enables one to develop what geologist Marcia Bjornerud calls "timefulness." Timefulness "includes a feeling for distances and proximities in the geography of deep time,"[1] allowing us to take in ways in which "the grandeur of Earth's story lies in the gradually unfolding, interwoven rhythms of its many movements," from the 500 million year Wilson cycle of supercontinent assembly and breakup, to the 50–100 million year lifespan of mountain ranges between orogeny and erosion, to the 1500 year Dansgaard-Oeschger cycle of rapid climate cooling and warming during the last glacial period in the Pleistocene.

One can set out in search of geologic timefulness through reading and research, but there is no substitute for sustained engagement with a geologist studying particular landscapes and rocks. I had the good fortune when I was living in Boulder to audit a course on the state's geology taught by Lon Abbott of the University of Colorado. In addition to being a widely published scientific scholar, Abbott is an enthusiastic popularizer of geology.[2] When I showed Lon the Poudre river stone, he identified it as an amphibolite, composed of hornblende metamorphosed in a basalt flow under high temperature and pressure. Our conversation and subsequent research into geologic maps of the area revealed that the amphibolite was likely formed at a depth of tens of kilometers in a shear zone around 1.7 billion years ago, during the birth of Colorado's continental crust. This was forged as a chain of volcanic islands like the Aleutians or Japan today, when present-day Colorado was at the edge of the North American plate. Because it formed at a tectonic plate boundary, its fabric of aligned minerals (dominated by hornblende) contains a light ribboning of plagiocase feldspar, foliation caused by the folding metamorphic environment.

On acquiring this geologic account of the stone, I felt a mixture of pride and humility—I was proud to have found such an incredibly old stone, formed deep in the Earth, and humbled by its age, which reduces one's own to

---

1 See Bjornerud (2018, 17).

2 With his wife and coauthor Teri Cook, Lon Abbott is the author of *Geology Underfoot in Colorado's Front Range* (Mountain Press, 2012), *Geology Underfoot in Northern Arizona* (Mountain Press, 2007), and *Hiking the Grand Canyon's Geology* (Mountaineers Books, 2003).

insignificance. The 1.7 billion-year age of the stone became a number to name, but with each iteration, its meaning seemed to elude me more. What does the age of the stone tell you about the stone's time? How is it possible to *think* this temporal scale, in any common or philosophical sense? As Stephen Jay Gould famously wrote, "An abstract, intellectual understanding of deep time comes easily enough—I know how many zeroes to place after the 10 when I mean billions. Getting it into the gut is quite another matter" (1987, 3). In a similar vein, John McPhee—perhaps the preeminent contemporary writer to grapple with geology—in his book *Annals of the Former World*, concludes that, "The human mind may not have evolved enough to be able to comprehend deep time. It may only be able to measure it" (2000, 90). The dichotomy McPhee draws between comprehending and measuring deep time opens a gap, perhaps leaving room for speculative approaches to contemplating deep time in different ways. I will consider two such ways in the remaining sections of this essay.

In response to McPhee's provocative remark that "the human mind may not have evolved enough to be able to comprehend deep time," one might take an inverse route. Perhaps it is not a question of humans having "evolved enough" to develop some higher cognitive capability, but a question of drawing on more ancient parts of our mind in contemplating deep time. Here, I am building on the paper I presented at the 2019 ISST conference, "Slow Time: Suspension of a Tension" (Harris 2021), in which I argued that "in order to reconcile our habitual human timescales with inhuman ["slow"] timescales, we need to practice a 'slow' mode of thinking." Essentially, this approach stipulates that "accommodating radically variant timescales is accomplished through 'the suspension of a [conceptual] tension'," which is accomplished in turn by "the suspension of [conscious] attention" (Harris 2021, 67). A "slow time" mode of thinking, in other words, provides a way to take in geologic and cosmic timescales, to hold them in thought or sense them intuitively, in ways that render them imaginatively (somewhat, somehow) comprehensible without necessarily comprehending them in a cognitive or analytic sense.

Time in general or deep time in particular are but one dimension or subject of slow thinking. Slow thinking, in turn, is but one in a set of practices I call absorbing stone, which encompasses collecting, displaying, photographing, researching, and thinking and writing about rocks. Absorbing stone is corporeal contemplation, incorporating a visual, cerebral, aesthetic, spiritual attentiveness and a tactile, tacit engagement—in the sense/sensing of "feeling stones in your bones." This relation constitutes a circuit or permeable membrane of being both absorbed by stone "without" and absorbing stone "within," a dual process of anthropomorphic animation of material balanced with geomorphic petrification of mind. Absorbing stone activates what Gaston Bachelard theorizes as the "material imagination," a form of imagination that "goes beyond the attractions of the imagination of forms; [it] thinks matter, dreams in it, lives in it, or, in other words, materializes the imaginary" (Bachelard 2002, xlvi). Like the circuit or dual process of absorbing stone, the work of the material imagination involves "a kind of dialectical animism" between matter and mind (2002, xlvi).

From a philosophical standpoint, absorbing stone yields what I term a 'stoned thinking',[3] a thinking with rather than about stone. In this zone of exchange, intermingling, or entanglement, mind becomes saturated with matter, at times merging with matter, at times suddenly distant. I often find my thoughts oscillating between moments of deep connection to stone and the resistance of rock to thought, its refusal to be anything other than sheer stuff. In those moments of bonding with stone, I experience what Paul Prudence, in an essay on Chinese scholar rocks, calls "the interanimating spirit of mind and matter where biological and mineral kingdoms collude, where their interpenetrating vectors are energized in a single field" (2020, 36). Implicitly subscribing to a kind of panpsychism, Prudence provides a neuroscientific account of stone contemplation: "(Meta) morphic resonance has opened a two-way transmission line so that geology can send messages in stone. The messages are mineralized into minds and in the marble patterns of neural pathways" (2020, 36). Prudence captures, and takes literally, the feeling when, in contemplating a stone, the contours of the stone absorbs the mind to the point that the contours of stone and of thought become indiscernible from one another.

The formal homology between lineations and textures on a rock and the intricate branching of the brain's neural pathways works well in the context of the wrinkled, tortured forms of Chinese Scholars' Rocks, such as Lingbi stones. But the river-smoothed surface of the amphibolite presents a more featureless, undifferentiated exterior. One's gaze is met with something like a blank stare. Trying to consider the stone's time, one encounters, to repeat Hyatt's resonant phrase, a "pool of silence holding a visceral stillness." Regarded this way, the stone holds a still pool of time that is, literally and figuratively, unfathomable. This impression recapitulates the notion of deep time as an "abyss," whose depths cannot be sounded, let alone circumscribed by the mind.

In order to take a measure of the stone's silence, I find it more productive to close my eyes and touch or hold the stone than to look at it. In his book *Listening to Stone*, Dan Snow, states that "Stone speaks through the hands because touch, being the oldest of our senses, is most sensitive to its language" (2008, xii).

3 See Harris (2018, 119–48).

One might say that touching stone lets us listen to silence. Or conversely, that listening to stone is a way to touch silence. Holding the amphibolite becomes a tactile encounter with stone and silence that opens a passage between human and geologic timescales, enabling not a comprehending of deep time but a tacit sensing of a specific stone's time. Kan Yasuda, a contemporary artist from Hokkaido who studied with Noguchi, asserts that, "Those who see and touch my sculptures, without dwelling on their appearance, will feel the flow of time. They will hear around them the resonating voices, yet silent, that emerge from the depth of the place. As if you are touching the accumulating of passed times that materializes in the present" (2007, 9). Listen as I might to the amphibolite, I have never heard "resonating voices." The dense rock remains dumb, maintaining a stony silence—an apt phrase evoking an active, willful silence. And so, in holding the stone, I feel it withholding its inner nature, holding its time within. I find such experiences of absorbing the amphibolite's time to be calming, in a manner consistent with contemporary discourses of mindfulness as a means of relieving stress and reducing anxiety. To measure the stone's time in this way is to conclude by dwelling in mystery, in an amorphous temporality which dissolves the difference between time and timelessness, geologic time and eternity.

In reflecting on these moments when time dissolves into timelessness, I realize that the *material* imagination has become an immaterial, quasi-religious imagination. I also recognize that I am thinking of the stone as an object and time as something that it holds. The age of the stone—1.7 billion years—has been extracted from Earth history and become a bloc of spatialized time, an infinite abyss contained in a finite thing. At such times one does well to recall Bjornerud's simple maxim that "Rocks are not nouns but verbs—visible evidence of processes" (2018, 8). Where the mystical abyss meets the material imagination, I find myself ascribing to a form of what Roger Caillois calls a "mystical materialism," or "an unknown and paradoxical kind of mysticism" (2018, 154). In recounting his contemplations of stones as a form of "spiritual exercise" (2018, 149), Caillois depicts memorable moments where, "between the stone's stillness and the mind's effervescence is established a kind of current where … I find wisdom and comfort" (2018, 154). While such moments of silence "dissolve in some inhuman immensity," Caillois stipulates that "that abyss would not be divine, it would even be all matter and only matter, active, turbulent matter of lava and fusion, earthquakes, orgasms and great tectonic ordeals; and motionless matter of the longest quietude" (2018, 154). Caillois's characterization here points up the ways in which it is easy to let knowledge of a stone's age turn time into a quantity measured by the geologic timescale, to let an unthinkable 1.7 billion years become an eternity. When a stone becomes an object containing "time," time is reduced merely to a timeline, a static metric along which events are strung as durationless moments.

In explaining the geologist's view of rocks as verbs, as "visible evidence of processes," Bjornerud turns to "the metaphor of a *palimpsest*" (2018, 22; original emphasis). Just as textual layers inscribe a palimpsest's history, she writes, "traces of earlier epochs persist in the contours of landforms and the rocks beneath" (2018, 22). In a similar vein, Cohen observes that "Stone conveys into present simultaneity the lingering traces of promiscuous relationships," including "mineral seams, fossils, enfolded strata, stripes of sediment involuted by tectonic force, alluvial deposits, inscriptions from rivers, from glaciers, from human hands" (2015, 98). Stones are archives, conveyors of their own history. We can flip the spatial metaphor of a rock as object containing time, and instead consider a rock as an ever-changing something that has emerged from time. Geologic time is indeed turbulent, as Caillois imagines it. Bjornerud's call for "timefulness" is a call to recall that the linear history recorded in the geologic timescale is marked by cycles of extinction and explosion, erosion and uplift. If we think of rocks as emerging from time, we realize that every encounter with a stone is a statistical anomaly, a meeting contingent on a rock's surfacing from the rock cycle at a particular time and place, where it is beheld or held by a human. This material fact renders the strange experience of feeling its attractive pull, such as the experience I had in finding the amphibolite, even more bewildering.

When I think of a stone emerging from time, literally surfacing from the depths of the rock cycle, I think of it as something pressed out by planetary processes: a stone as an 'expression' of Earth. This notion is evoked poetically by Louise Erdrich in her short story, "The Stone". Shifting the narrative point of view from the human protagonist to the titular stone of the tale, Erdrich writes: "A stone is in its own way a living thing. Not a biological being but one with a history far beyond our capacity to understand or even imagine. A stone is a thought that the earth develops over inhuman time" (2019, 44). In thinking *with* stone rather than about stone, then, we might consider a stone as a geologic expression, "a thought that the earth develops over inhuman time." But rather than simply slipping into an anthropomorphic animation or personification of stone, we might rather use Erdrich's phrase as an invitation to think of a geomorphic petrification of thought. That is, we might conceptualize thought—human or terrestrial—as a material, recursive process. The rock cycle exemplifies the Earth's self-involvement, its constant remixing of its own contents, its transformation of internal materials (with the occasional

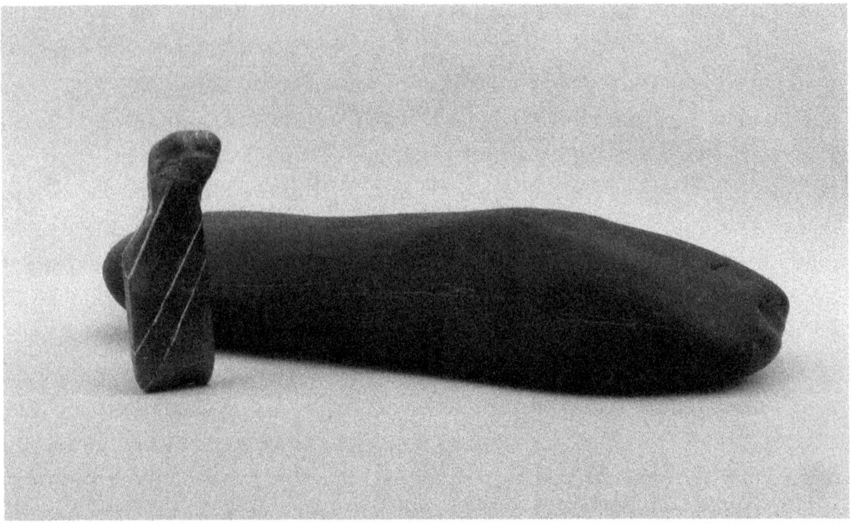

FIGURE 4.5 My Geomorphic Mind. Dreams of Stone (2023)
PHOTO BY PAUL HARRIS

bombardment from space). Here, taking measure of a stone's time occurs in a kind of zone of exchange, where human thought and earth thought intermingle. In thinking along these lines, I dream a lithic avatar of myself contemplating the amphibolite.

What, then, happens when I try to think of taking the measure of the amphibolite's time by considering what I can reconstruct of the history from which it emerged? Metamorphosed under tremendous heat and pressure in the Earth's crust at the boundary of a tectonic plate, it was ejected—expressed from and by the Earth—in a volcanic shear zone. In recent times, it has clearly been moved and smoothed in the Cache de la Poudre river, as the river sculpted its course through the canyon. Between the amphibolite's formation and its contemporary history, it has persisted through two cycles of mountain-building. It can be conjectured as being present when the Ancestral Rocky Mountains were uplifted during the Pennsylvanian period 323–298 million years ago, at the time of the supercontinent Pangea. It can be imagined as enduring the process as the Ancestral Rockies eroded completely and a shallow continental sea took their place during the late Cretaceous period (95–65 million years ago). It was likely present as the current Rocky Mountains uplifted during the Laramide Orogeny (65–40 million years ago, late Cretaceous into Paleogene periods).

How might we think about or try to comprehend these cycles of uplift and erosion? In *A Land* (1951), her singular geologic memoir, the archaeologist and writer Jacquetta Hawkes thought of these cycles as a terrestrial respiratory cycle: "The history of the earth's crust has a rhythm. Denudation weakens it, the mountains are rucked up and the molten layer below forces itself toward the surface, then the storm dies away and denudation begins again. If the movement were speeded up, as in a cinematograph, we should see a rise and fall as though of breathing" (Hawkes 2012, 15). Similar images of geo-respiration, the living breath of a living planet, can be found in many Daoist works. In *The Great Image Has No Form*, Francois Jullien offers a condensed view of this ontology: "Everything in the universe originates in the same breath-energy. … It leads to the infinite diversity of beings, man included, and to their relations with one another and their cohesion within a landscape … rising and falling, soaring upward or sitting down, the mountain brings about the great respiration of the world" (2009, 135).

In light of this comparison, in taking the measure of the amphibolite's time, I sometimes try to calibrate my breathing with terrestrial respiration. I initially considered this calibration in quantitative terms: 'if each breath I take corresponds to a million geologic years, then. …' But I discovered a more intuitively satisfying approach by once again turning to Bachelard. In *Air and Dreams*, Bachelard invites us to "listen to nothing but our own breathing," which enables our material imagination "to become as aerial as our breath …" (1988, 241). Bachelard's close readings of poetic images is informed by a version of mimology, the notion that words—their sounds and shapes, lengths and patterns—imitate the world. This outlook is seen when he loses himself in reveries (a key Bachelard term) of words. In *The Poetics of Space*, Bachelard's aerial imagination and mimology converge in a fascinating manner:

> The word vast, then, is a vocable of breath. It is placed on our breathing, which must be slow and calm. … It expresses a vital, intimate conviction. It transmits to our ears the echo of the secret recesses of our being. … With it, we take infinity into our lungs, and through it, we breathe cosmically, far from human anguish. … For it is a word that brings calm and unity; it opens up unlimited space. It also teaches us to breathe with the air that rests on the horizon. …
>
> BACHELARD 1964, 196–97

I have used this passage as the basis for a thought-experiment or thought-exercise, or what I think of as an "Imagination Stretch." Holding the amphibolite while standing with feet shoulder width apart, I whisper the word vast as I inhale deeply while sinking down and bending my knees, then exhale slowly and rise calmly back to standing position. In this way, I feel in my body some vestige of the stone's history, even if in a simplified manner. The combination of the physical exertion of holding the stone with the meditative mindset provides a different, fruitful attempt to take the measure of this stone's time.

The attempt to measure a stone's time, or to take its measure, has proved a productive process of experimentation. Through these different ways of trying to understand its time, I have deepened my relations with this specific stone. But while I have deepened my feeling for the deep time of the amphibolite, this in no sense means that I comprehend its time. To measure its age is not to take the measure of its time. In this spirit, the most appropriate way to end this essay is to cite the final sentence of Francis Ponge's marvelous prose poem "The Pebble": "He attempted to describe stone and petered out" (2003, 99).

References

Aisekikai. n.d. "The Suiseki tradtion." Accessed February 10, 2025. https://www.aisekikai.com/the-suiseki-tradition.

Bachelard, Gaston. 1964. *The Poetics of Space*. Translated by Maria Jolas. New York: Orion Press.

Bachelard, Gaston. 1988. *Air and Dreams: an essay on the imagination of movement*. Translated by Edith R. Farrell and C. Frederick Farrell. Dallas: Dallas Institute of Humanities and Culture.

Bachelard, Gaston. 2002. *Earth and Reveries of Will: An Essay on the Imagination of Matter*. Translated by Kenneth Haltman. Foreword by Joanne Stroud. Dallas: Dallas Institute of Humanities and Culture.

Bjornerud, Marcia. *Timefulness: How Thinking Like a Geologist Can Help Save the World*. Princeton University Press, 2018.

Caillois, Roger. 2018. "Excerpts from *Pierres reflechies*." *Rock Records: Special Issue of SubStance* 47 (2): 49–155.

Cohen, Jeffrey Jerome. 2015. *Stone: An Ecology of the Inhuman*. Minneapolis and London: University of Minnesota Press.

Cohen, Jeffrey Jerome. 2018. "Feeling Stone." *Rock Records: Special Issue of SubStance* 47 (2): 23–35.

Erdrich, Louise. 2019. "The Stone." *The New Yorker*, September 9.

Gould, Stephen Jay. 1987. *Time's Arrow, Time's Cycle*. Cambridge: Harvard University Press.

Harris, Paul. 2018. "Stoned Thinking: The Petriverse of Pierre Jardin." *SubStance* 47 (2): 119–48.

Harris, Paul. 2021. "Slow Time: The Suspension of a Tension." In *Time in Variance*. The Study of Time 17, edited by Paul Harris, Arkadiusz Misztal, Jo Alyson Parker, 66–93. Leiden: Brill.

Hawkes, Jacquetta. 2012. *A Land*. London: Collins Nature Library.

Hyatt, Carl Austin. 2021. "Listening to Stone Beings." *E-flux journal*, 1 July. https://www.e-flux.com/architecture/survivance/402520/listening-to-stone-beings.

Jencks, Charles. 2005. *The Garden of Cosmic Speculation*. London: Frances Lincoln.

Jullien, Francois. 2009. *The Great Image Has No Form, or On the Nonobject through Painting*. Translated by Jane Marie Todd. Chicago: University of Chicago Press.

McPhee, John. 2000. *Annals of the Former World*. New York: Farrar, Straus and Giroux.

Ponge, Francis. 2003. *Selected Poems*. Translated by C. K. Williams, John Montague, and Margaret Guiton. Winston-Salem. Wake Forest University Press.

Prudence, Paul. 2020. "Scholar's Rocks: The Interanimating Spirit of Mind and Matter." *Reliquiae* 8 (1): 25–37.

Snow, Dan. 2008. *Listening to Stone*. Photographs by Peter Mauss. New York: Artisan Books.

Yasuda, Kan. 2007. *Touching the Time*. Milan: Skira Editions.

CHAPTER 5

Logogenesis: the Genesis of Laws in the Postmodern Vision of Time

Frederick Turner

Abstract

The spirit of postmodern discovery involves two main moves away from classical modernist paradigms. The first is a move from abstract principles to statistical and emergent systems in which we no longer see the world as objects governed by principles but rather as principles generated by objects. The second is a move from measurement as the application of a theoretical measuring stick to measurement as a comparison between two quantities, each deriving from their context, and thus partly from each other. Examples in physics include relativity theory and quantum mechanics, and appear in the social sciences in the move from planned to unplanned but regulated economies, from legal resolution of moral disputes to negotiated settlements, and from central bank currencies to blockchain-based currencies. Redefining the act of measuring in these postmodern terms suggests that space is measured not by units on a grid but by the number of objects, lying between one point-event and another, that force the traveler to take a turn before arrival. Applying this concept to the measurement of time, we find that we are reckoning the number of present-events, or turns, that must be dealt with between two points whose separation we want to know. Time is now not an abstraction but the actual *difficulty* of the passage, as the number of possible itineraries mount up, as in the traveling-salesman problem. The paper concludes with a meditation on the meaning of the word “verse” and the use of the “turn” in the “numbers” and “measures” of human poetry.

Keywords

entropy – measurement – emergent systems – relativity – quantum theory – postmodern – acceleration of change – information theory – turns

 | DOI:10.1163/9789004723634_007

1 Framing the Question

The spirit of postmodern discovery involves two main moves away from classical modernist paradigms, following the loss of classical benchmarks. The first is a move in the basis of explanation: from abstract principles to statistical emergent systems. Space is no longer Newton's eternal invisible perfect three-dimensional grid, but a wild changing landscape of pits and plateaus and vortexes, differing according to where one stands. Time does not just accelerate and dawdle, but attenuates and thickens, while some local times serve as measuring-sticks for the antics of their more volatile neighbors and for the wholes of which they are parts. Wholes reciprocally determine the behavior of the parts that brought them into being. The laws of science were not eternally given, but emerged from events. In that spirit we no longer see the world as objects moving in time and space and governed by principles, but as a single entangled quantum spacetime manifold whose evolving decoherence generates new forms of energy and matter and ad hoc rules and frames to allow their coexistence.

Examples in physics are clear in relativity theory, in which a mass, itself a knot of energy, partly determines its own reference frame of space and time. In quantum mechanics, distinct quantities of mass and charge and motion are the result of the mutual adjustment of entangled harmonics and their decoherence into locality and identity by interaction and mutual observation. Later in this chapter we will look at the implications of this twentieth-century discovery as they have played out more recently.

Even in mathematics itself, the essential "go-to" of all the other scientific disciplines, the idea of fixed, timeless, and eternal principles, is under assault. In the field of algorithms, for instance, the scale of difficulty of solution—now mathematically definable—mimics the characteristics of physical time. Problems are solvable within a short, long, or unlimited "time."[1] Let us compare some algorithms, ranging from not difficult at all to very difficult. If one is asked what number is five, the answer is "five"—it takes no time to accomplish the algorithm. If asked what is two plus three, the algorithm would take perhaps three logical operations to get the answer (add one, repeat, repeat). Now take the "traveling salesman" problem—what is the shortest route that includes three cities? Four? Ten? The number of operations required to perform the algorithm for solving the problem increases nearly exponentially ("non-polynomially," or "NP") with the number of cities. At ten cities, no

1 For a good summary of the issues, see "Computational complexity theory" (2025).

ordinary computer can efficiently solve the problem in a reasonable time. At fifteen, the number of possible routes that would have to be checked is 1,307,674,368,000, according to Wikipedia. Even the most perfect computer, whose components would be all the matter in the universe, must take a certain amount of time to perform an operation, limited by the specific minimum chronon of its universe. At, say, one thousand cities, there might not be enough time in any conceivable universe, and the entropic heat generated by the process would burn up the computer.[2] (Indeed, far from being an example of the perfection of the timeless, mathematics might be a sound basis for the definition of thermodynamic time itself.)

Taking a related but different case, mathematics, once thought of as pure and timeless, cannot avoid the problem of ordering—firstness in any algebraic or logical statement is not the same as secondness, and commutative mathematics differs from noncommutative mathematics.[3] The distinction itself is the germ of time. Mathematics in practice is an ongoing physical process, depending on the kind of computational systems that have evolved at a given time. Kurt Gödel's incompleteness theorem dictates that any kind of unlimited recursiveness such as that embodied in the statement "This statement is unprovable" renders void the idea of a primary eternal principle.

The second change is a move away from the idea of measurement as the application of a measuring stick to a theoretically ascertainable quantity, towards measurement as a comparison between two or more quantities, each acting as a measuring stick for the other, each conditioned by its shared and unshared context and helping define the context itself. The observer effect, first noticed by early quantum physicists such as Heisenberg, Bohr, and Schrödinger, is not a problem or mistake in our ability to measure, but a fundamental characteristic of the universe—not a bug but a feature. Measurement is an act of knowledge, and the state of knowledge of a system is as much a part of it as its atoms and waves.[4] Here disagreement, faultiness and difficulty are systemic; events *are* the mutual adjustment of the measures, accuracy a function of use and application. Given these two changes, from principles to emergent systems, and from one-way abstract measurement to mutual concrete measurement, how are we to properly measure time?

2 Rolf Landauer, a researcher in the thermodynamics of information processing, made this point to me at a meeting I helped organize at the University of Texas-Dallas in the 1990s. Quantum computation is an attempt to overcome the limitations of any sequential form of computation, but the fragility of entanglement, upon which quantum computing relies, looks like another version of the same problem.

3 See Heller and Sasin (1998).

4 See Ananthaswamy (2017).

2 The Evolution of Measure

Twentieth century quantum theory has proposed that the most fundamental level of physical reality is probabilistic in its nature, and that observation could force a constitutively indeterminate field of possibilities to manifest itself as a distinct and local object such as a subatomic particle. To observe means fundamentally to measure, that is, to characterize what is observed according to the observer's yardsticks. But to do so changes not only the observer's state of knowledge, but also the physical reality of what is observed. More recently, theoretical physicists have established that a human or even a living observer (such as Schrödinger's cat) is not needed to perform this transfer from the world of possibility to the world of classical reality, a transfer now termed "decoherence." Any constituent part of the universe may perform this role of rendering definite what was, beforehand, essentially ambiguous.

But this cosmic process of mutual observation, measurement, or decoherence is not just a chaos of conflicting versions of reality. More recently, theoretical physicists such as Wojciech Zurek, Carlo Rovelli, and Christopher Fuchs have begun to establish emergent processes that resemble the mutation-selection-heredity cycle well known in biological evolution. These processes account for how the highly differentiated and robust reality we see around us emerges from indeterminacy. Variously characterized as "Quantum Darwinism," "Quantum Bayesianism" (after the prediction system of the mathematician Thomas Bayes), and "relational quantum mechanics," these approaches suggest that the emergence of agency in humans and living organisms is not a freak in a meaningless universe but a result of a natural teleogenic—or "meaning-making"—tendency embedded in time (the process of the universe) itself. Natural laws are hammered out by mutual measurement, ensuring the continuance of the agents of measure.[5]

Before the breakup of the Superforce into gravity, the weak and strong forces, and electromagnetism, there were no laws of physics. Before the elements emerged out of the early universe there were no laws of chemistry. No such thing as electron shells, valence, compounds, polymers, etc. Before carbon came along, with its emergent benzene ring, there were no principles of organic chemistry or biology. The emergence predates the scientific laws it creates that now "feel" as if they were eternal. Chemistry is a discipline in which catalysis, a paradigmatically whole-to-part process, is the origin of almost all new kinds of material substances. Autocatalysis (think crystallization, for instance) maintains and extends all enduring wholes, within which individual

5 See Azarian (2022, 268–74) for a very good and accessible account of this burgeoning field.

atoms are at the command of the systems they consist of. Ecology is the science of organically "made-up" regimes of order, with their own emerged laws and principles. An economy or society is a principle-generating system, understandable only through the principles that it itself generates.

Biological evolution is a burgeoning of clocks. As J. T. Fraser showed,[6] the human body contains a multitude of clocks whose basic pendulums range from the relatively generic vibrations of atoms, through the more complex rhythms of autocatalytic chemical processes, the metabolic cycles of the various organs, and the vast range of neural rhythms, to the social and technological periodicities the body has internalized. Each of these clocks is a measure of the others. They regulate each other by competition and collusion in a sort of market or deliberative debate, which, when it breaks down, constitutes illness. Metabolism may be viewed as the mutual correction of the clocks. Desire is the gradient toward synchrony; action is possible only in the presence of contrary measurements. Hunger is a force that attempts to eliminate itself. Evolution adds new clocks to the mix to settle the disputes and environmental challenges of the old. On a larger scale, ecosystems have their own collections of clocks, more or less synchronized or syncopated, and species emerge from the adaptive pressures that mismatches in rhythm create. In a sense, time is not so much something we all live in, as something that we all *do*.

In the social sciences, there has been a similar shift, from a view of governing principles ruling passive objects to one of emergent systems arising from interactive agents, and from standard measures to changing ones. We have shifted from planned to unplanned but regulated economies, from the legal resolution of moral disputes to negotiated settlements, and from central bank currencies to mutually dependent currencies. Top-down command politics and planned economies have shown their brittleness and inefficiency; messy democracies and unplanned but consensually regulated economies seem to work better. Negotiated settlements seem to have better outcomes than legal absolutes, in which positions are hardened according to principle. Again, what was a bug—the argy-bargy of haggling and debate—is a *feature* of successful societies. The argy-bargy is the chiming of different clocks as they search for compromise—and make discoveries along the way.

We need a term that describes the emergence, not of things and processes in themselves, but of the new rules, principles, and forms of measurement that emerge with them and regulate them. A new logic or logos comes into being with a new order of matter, life, and mind: perhaps "logogenesis" may serve our turn.

---

6 See Fraser (1999).

3 The Acceleration of Logogenesis

Putting together the two postmodern moves—from the primacy of principles to the primacy of interaction, and from absolute measures to mutual dynamic measures, we have come to see the world as change, and not just change but continually faster change. If we look at the history of the evolution of this planet, we notice one overwhelming phenomenon: *acceleration* of change. Simple geology gives way to the more complex chemistry of plate tectonics and the much faster systems of evolving life, with huge boosts to the rate of change from the anaerobic to the aerobic, from unicellular to multicellular, from asexual to sexual reproduction, from monocultures to many-layered ecologies, from primary producers to predators, and from relatively asocial animals to communities of highly encephalized individuals that take loving care of their young and thus presuppose a future that can be partly predicted and created.

Genetic evolution requires mutation to generate new combinations of genes that may enhance survival and facilitate reproduction. Relying on mere random damage to alter gene sequences is very wasteful and inefficient: bacteria and archaea developed ways to swap good working gene sequences; sexual reproduction enabled a controlled process of sharing beneficial mutations; and epigenetic variation accelerated the rate of change. A pool of individuals function better than a pool of clones. As genes developed more and safer forms of mutation, evolution and speciation accelerated, new ecological niches appeared, and new avenues of logogenesis opened up. When Homo sapiens appeared, and the Anthropocene began, logogenesis left genetic invention behind to follow culture, and flowered into a bewildering variety of social and linguistic branches.

Human history's major developments used to happen over a scale of hundreds of thousands of years. But as human societies multiplied and became more sophisticated, significant changes started to take place over tens of thousands of years, then thousands, then hundreds, and now tens. But the earlier time-measures did not go away: they serve as the yardsticks for the later, more densely temporalized later ones. We measure our electronic technology in nanoseconds, our mechanics in microseconds, our cookery in minutes, our animal husbandry in years, our changes of political governance in decades, our analysis of history in centuries Each of these measures has its own tone and flavor, and together they constitute a rich productive marketplace of discovery and emergent order.

The universe is not a thing measurable by a single measuring-stick, but a multiplying body of mutual measuring-sticks, the later ones emergent from and partly controlling the older ones. The "clocking" of more complex living organisms uses a range of measures—diurnal, monthly, annual—even millennial if we consider the response-time of a genome to environmental change—to better predict and thus establish its future. An advanced composite clock, such as a social animal, is able to manipulate the more primitive clocks of inert matter or chemical reaction. We humans use the more advanced clocks of such animals to feed and clothe ourselves.

The competitive element of evolution is not only one in which species fight it out for the domination of an ecological niche, but also one in which species compete in the creation of new cooperative niches. The competition can be described as a predicting contest, in which each competing species maps out a future hospitable to its own potential and acts as if that future were coming. The opposite of competition—commensality, symbiosis, or trading—is a matching of compatible logogenetic maps, as happens for instance between the flowering plants of this planet and their insect pollinators.

This picture of radical and systemic acceleration may also render meaningless the gloomy cultural concern about the end of the universe that has marked the last two or three centuries. If the number of significant events per physical second is increasing exponentially, and with it the technology for experiencing them, then the volume of *experienced* time will have become infinite for all intents and purposes long before the imagined heat-death of the physical world.

4 A Critique of the "t" Axis

In any case, we are talking now of measuring sticks that vary according to what they measure, including their subject's own measuring system. To extend the metaphor of the "measuring stick" a little more, let us consider the "t" axis, the time-line, as found in every modern discipline from math to political philosophy, that measures things and processes. If we dwell in a contemporary civilization, our science, our social and technological organization, our everyday ordering of events and actions are all based on the validity of mapping the passage of time onto a single spatial metaphor. If we unwind the clock-face, we get an infinite line (since 12 in the evening is also 12 in the morning). If we don't turn at the end of the calendar's graphically-imposed furrow, whether weekly or monthly, we also get an infinite line.

But is that linear "t" axis really a legitimate assumption? Since the time of graduated candles, clepsydras, sundials and stellar timekeeping, we have based our time measurements on a marked line, using for this purpose the spatial characteristics of physical objects and processes such as burning wax, moving stars, or flowing water. As Bergson insisted, we have simply mapped time onto space. Natural recurrences, such as the breath, the heartbeat, the rising and setting of the sun and stars and the phases of the moon, are taken to be of equal temporal length. We ensure the accuracy of our longer units of time by standardizing the number of shorter units they contain, giving us, for instance, the Babylonian three hundred and sixty degrees in the circle of the sun, twenty-four hours per day, and sixty seconds per minute. But how do we know that one day is as long as another, unless we count the hours in each?—and how do we ensure each of those hours are the same except by identifying the minutes they contain and counting them? In regard to space we can at least use a compass or caliper, rotating it repeatedly along the line and counting the rotations. But there is no temporal caliper, there are no second and third dimensions of time to accommodate a free foot.

In search of greater accuracy we propose shorter and shorter time-intervals and invent finer and finer machines for recognizing and counting them. Today we have narrowed the interval and increased the frequency of the presumably regular oscillations of the processes we use to calibrate our scale. We are down to the 247 zeptoseconds[7] it takes for a light particle to traverse a hydrogen molecule. But how can we check whether two measured zeptoseconds are exactly equal to each other? And what use would such a measure be? To use that zeptosecond interval, as ancient in itself as the Big Bang, to describe a biological or social event, would be as absurd as to try to get the meaning of this sentence by breaking down the printed letters into their nanometer pixels. Although everything emerged from the first zeptosecond of the Big Bang, that zepto-stick is now irrelevant to almost everything that really happens.

The mistake, perhaps, is to try to separate the realm of knowledge from the realm of physical reality. The cycles, rhythms and oscillations of the animate and inanimate worlds cannot be measured against an ideal mathematical *chora* (in Greek χώρα, an archetype of a unit of space), but only against and in terms of each other. The true meaning of Pythagoras's equation of music with geometry is not, perhaps, that music arises from geometry, but that geometry arises out of the music of nature. The world is temporally regulated not by a single cosmic autocratic rule-giver, but by the robust but continually self-adjusting democratic consensus of all its constituent parts, weighted according to their relative powers of prediction and degree of local nestedness.[8]

7 See Grundmann et al. (2020, 339).

8 A human or a cat is more determinative of the near future than is a single yeast cell, because the cat and the human contain, nested, vast numbers of cells, each individually no more complex than a yeast cell but "voting" together.

Saint Augustine famously distinguished shorter and longer time-intervals in terms of the long and short syllables of Latin poetic meter. Poetry did not begin to exist once we had clocks that could measure the fraction of a second a syllable takes up—and music existed, with exquisite perceptible nuances of timing and tone, long before Heinrich Herz's name was given to the scale of wave frequencies. Nature, and human natural perception, is already actively measuring itself against its surrounding and internal environment of rhythms and cycles, and the measuring is itself an essential component of its material existence.

As the universe has evolved, especially on this planet, it has been increasingly governed by the clocks of chemistry and life and the technologies they created. The density of significant changes (as opposed to mere stasis, cyclic repetition, or chaotic violence) increases with time. More new stuff gets packed into each successive millennium, century, year and day. This is partly because previous forms of change continue while also being subsumed into new forms of change, such as ecosystemic adaptation, sociobiological evolution, economic development, sociopolitical interaction, technological innovation, and cultural invention. And the older measures often act as actuating irritants to spur on the function of the new.

The linear spatial t axis is useful as a base to measure the raw dimensions of processes that began long after the established rhythms of electronic or planetary orbits. But in the realms of life and culture other regularities, such as species life cycles, metabolic cycles, climate cycles, economic and historical cycles, and even musical or poetic meters, may make better baselines. Depending on the questions one wants answered, these baselines may be more inclusive of the actual complexities of reality. We can go on using the t axis, but only if it is specifically defined in terms of other measures.

5 A Logogenetic Description of Space and Time

One upshot or takeaway of this examination of the practice of measurement is that it is the contents of what is measured that dictate the form of measure. There is no prior container or vessel in either space or time. Empty space is no space at all. Empty time is timeless. A photon in a total vacuum is perfectly still and instantaneous; it is noplace because it has nowhere to go and nothing to go past. And time and space are not just symmetrical at this point but indistinguishable. We need a term for the kind of difference or content that would be required to give our photon a place to be and a time to get where it is going.

There exists such a term in English: the word "turn" or "turning." Spatially we direct a stranger by a sequence of turns or turnings—"second on the right, then third left."

Temporally we take turns in a game (Heraclitus claimed that time was a game[9]), and if the game represents a real-world action, as does a war-game, there is a difference between the player's time and the time of the events depicted in the game.[10] But given the large element of gameplay in our real life, and the usefulness of games in real life, there is a certain bleeding between fiction and reality: the ease and quickness of the game-move makes us impatient when we want to make the same move in the real world. We grow quickly bored with a game that does not offer time-consuming tactical or strategic puzzles, puzzles we would not have the time or patience to solve in real life.

In mathematics, as we have already noticed in the case of commutation, if we "turn" one kind of equation backward, we get the same answer; if we "turn" another kind backward, we get the opposite answer. Thus the fundamental content that creates the container of any event or object is the set of turns or turnings it contains. Time and space are "made up" or constructed by the turns it takes each of its constituents to get from one place to another, from one moment to another. Space, as we know from current cosmology, is expanding as more and more "turns" are needed. And time, as we know from Fraser, adds on new features with every evolutionary emergence, packing more and more "turns" into each old-style second.

So the governing properties of complex systems do not pre-exist those systems, but emerge from the interactions of the agents which compose them. Perhaps we need measures that actually reflect this reality. Euclid still holds in Euclid's world, but this world is not that one alone; it is now encrusted with nearly fourteen billion years of more and more messy measures.

Space itself is not homogeneous, as we noted at the beginning of this essay. A context rich with stuff, like an English garden, contains a lot of space. Interstellar space contracts up to the limit of the speed of light.

The measures that spacetime's own contents apply to their surroundings are themselves the encrustation that force the traveling observer to make a turn, to take into account the local rules. Taking a turn is the most fundamental unit of time. In this sense space is measured not by units on an infinite

---

9 See Arnold (2018).

10 Laurence Sterne's (2023) *Tristram Shandy* famously plays with this mismatch.

eternal abstract grid but by the number of objects or events that lie between one point-event and another. These are the objects and events that "get in the way" of traveling from one point-event to the other and so force the traveler to take a "turn" in *time* and deal with them before continuing. The classic example of such a turning-place would be a nearby black hole, which massively distorts both the view and the trajectory of the traveler, both the measuring-stick and the clock. But a humbler example may be more vivid.

When we calculate a car trip, we know that to pass through the ten miles of the crowded city will take more time than the fifty freeway miles that follow. The fifty miles are a relative abstraction; nothing much happens there except for the occasional billboard. In contrast, negotiating the objects and people in the city is a complicated and thus concrete reality. The Earth is a city, and it is much bigger in this sense than the solar system, which can be traversed in a few hours by a sufficiently fast chunk of rock. To take the city fully into account, we would need a massive effort of information-gathering, which itself would impede our passage and prolong the time—in multiple ways. Space would crystallize into time. And the very work of learning the city would itself become part of the business of the city, an addition to its crust of unique information. What makes the Earth a city is the enormous body of measurement that all its parts and systems have accumulated to measure each other.

6 Information and Entropy: the Logogenetic Turn

We measure to reduce our uncertainty, and to do so we must reduce what we measure to those elements that are deterministic, that is, that either remain the same over time or whose past and future can be calculated from their present state with the minimum of error. We can only get through the city quickly by massively increasing our ignorance of our surroundings, by ignoring all the elements of it that are unique, temporally asymmetrical, and unstandardized. We obey the traffic signs, the unambiguous measures of the physics of the roadbed, and the simple conventions that govern whose turn it is to move. The city seems disordered to us, but that is because of our desire for certainty. In other words, certainty involves not the increase of information but its decrease. The more ignorant we are, the more secure we are in our measurement. And here we run into what now looks like a gigantic historical error, involving entropy. This is an error that J. T. Fraser, our founder, saw only too well, but that has only begun to be recognized now that information theory has begun to make it clear.

The traditional modernist description of time[11] linked it to the increase of disorder, defined by the statistical mechanics of the nineteenth century as entropy, as in the work of Maxwell, Boltzmann, and Josiah Gibbs. This vision of the world of time was radically pessimistic. We are all fighting over a limited and diminishing stock of usable work energy, identified as order, and laying it waste as we do so. Even before its full moral implications were explored by the existentialists, we find an expression of this pessimism in Tennyson's *Idylls of the King* and Arnold's "Dover Beach":

> the world, which seems
> To lie before us like a land of dreams,
> So various, so beautiful, so new,
> Hath really neither joy, nor love, nor light,
> Nor certitude, nor peace, nor help for pain;
> And we are here as on a darkling plain
> Swept with confused alarms of struggle and flight,
> Where ignorant armies clash by night.

In the same period, however, Darwin had demonstrated an enormous counter-example to this whole vision of things: the evolution of life on the Earth, its staggering profusion, its ongoing multiplication of exquisitely complex and ordered forms. But this radical counter-example was dismissed as a local blip, a chance, a brief swerve in the universal decay. It is not germane here to speculate why the educated opinion of the time made so perverse a choice, but only now are we beginning to understand the nature of the mistake.

The brilliant insights of Claude Shannon, John von Neumann and others have finally begun to correct the picture. As it turns out, the mathematical description of information in the terms of information theory is identical to the mathematical description of entropy in thermodynamics. Entropy is what happens when a system explores the information-space of its various possible futures—that is, when it *creates* information. It thus evades our desire for certainty. We cannot know from its measurement what something was in the past and will be in the future. In other words, the increase of thermodynamic entropy over time is essentially the increase of information, that is, of *order that we do not (yet) know*. As we know from our example of negotiating the way through the city, we must increase our ignorance to be certain enough to

11 Though Lucretius anticipated the idea two thousand years ago in his *De Rerum Natura*.

find our way through. We were confusing the multiplication of information with disorder, we were mistaking our method of understanding-by-reduction for the reduction of order by decay. In other words, the universe is not running down but flourishing and growing into wilder and wilder realms of order, which our measuring struggles to stay ahead of.

The modernist picture of time as the entropic exhaustion of a limited stock of order and agency has been turned on its head.[12] Our model of measurement needs to change, from passive extraction of information into the recognition of informational order and agency, and the activation and emergence of new information through that recognition. Enlightenment science, which measured things in seconds and meters, was struck with awe and wonder at the vastness of the cosmos and the tiny insignificance of the human world; but if we use as a measure the relative density of turnings in space and time required to constitute a biological, social, or mental event, it is intergalactic space that is tiny, and a human birthday party that is enormous.

7 Postscript

In my epic poem *Genesis*, I describe how the first humans to set foot on Mars experience the difference between Earth-time and the time of Mars. The poem attempts to convey by physical detail the shock of temporal absence: the absence of the rich broth of time that embraces us everywhere on Earth, that nourishes our life and sustains our reason, that Mars either never possessed or possessed once but almost utterly lost. But there is a beauty and a pathos in that maimed world that I painfully expressed in the old medical word for Down Syndrome: "mongoloid." (This was the term used by the compassionate doctors for my baby sister Lucy, whom we dearly loved but who, crippled by a dozen physical and mental deficits, passed away in 1958 at five months old.) The task of the traditional human epic poet (including not only the verbal artist but also the prophet, shaman, inventor, and scientist) was to refine the Earth's rich inheritance of temporal fertility, to create desire, the fuel of culture. But the task of the Martian poet (here embodied in my character Beatrice, the planetary gardener, who has had a dream about the Cumaean Sybil), is to begin to create from scratch that rich milieu, that productive broth,

12 Azarian's very useful book (2022) gives a good and comprehensive account of this recent reversal in our understanding of entropy. Other less radical proposals exist to solve the creation of novelty problem in the Universe, see for instance Siegel (2024).

in an existential wasteland. Poetry in this wider sense, the poem claims, is the most potent agent of measurement/creation that has yet emerged.

First, though, let us recall how it had been
Before the helmed conquistadors had come;
What lay below the gold wings of the ship
That bore the speechless astronauts to ground.

A numb plain spread with stones. A weary steppe
All bleached to tired red with ultraviolet.
Soil crusted, sere; limonite, siderite.
Hard radiation in a waste of cold.
Rocks sucked dry by the near vacuum.
Stunned with the blank math of the albedo
The eye tries to make order of it, fails.
Whatever's here once fell from someplace else.
Sometimes a crag a foot high, or a mile;
Always the sagging tables of the craters,
The precise record of a mere collision.
And yet a stunted and abortive chemistry,
A backward travesty of life, proceeds:
Parched cirrus clouds move over the ejecta;
A hoarfrost forms upon the shadow sides;
Dark patches colonize the regolith;
Sometimes with a thin violence a sandstorm
Briefly makes shrieks of sound between the stones;
Rasps off their waists and edges, and falls silent.
Time here is cheap. A billion years can pass
Almost without a marker; if you bought
A century of Marstime in the scrip
And currency of Earth, you'd pay an hour
Or half an hour of cashable event.
It's really a young planet then, a bald
And mild mongoloid, a poor old cretin
Worth but a handful of Earth's golden summers.

And it was beautiful. Those who first walked there
Said it was fresh as the true feel of death,
As Kyoto earthen teaware, as the Outback;
As clean as is geometry, as bones.

To spoil this archetype, this innocence,
Was to incur a guilt whose only ease
Was beauty overwhelming to the loss,
Was a millennial drunkenness of life
That might forget its crime in ecstasy.
But it was not enough to reenact
The long sensualities of mother Earth,
To take on trust the roots of history;
They must be minatory, and exact
A last accounting of the failed balance
All the intestate dowager had left.
There must be new assumptions in the matrix,
New ratios, dimensions and arrays.

***

The poets of the Earth refine the fuel—
The hot benzene of value culture burns
To power its subtle engines of desire—
From fossil liquors buried in the stone
Through ages of creation and decay.
They can afford to toss aside the raw,
And take for granted a world cooked and rich
With ancient custom, languages numberless
As layers of autumn leaves within the forest,
Nature itself grown conscious, turned upon
Itself to make its rings so intricate,
History fertile with its own grave-mould.
And so the poets' work is little more
Than cracking out the spirit they inherit
In the tall silver towers of poetry
To brew those essences, those volatiles,
Those aromatic esters, metaphor,
Image, trope and fugitive allusion.
Prodigals, they burn half that they use
To purify the rest; and they make little,
Only a froth or lace of ornament.
The poets of Mars must brew the very stuff
The Earth-poets burn as waste; must mate each word
With breeder's care, and dust the yellow pollen

Over the chosen stamen; graft the stem
To coarser stock, and train the line to sprout
Productive variation years ahead.
The poets of Mars must make the myths from scratch,
Invent the tunes, the jokes, the references;
They must be athletes of the dream, masters
Of the technology of inventive sleep,
Architects of the essential shades of mood.
What they inherit from the Earth, they earn,
Through sacrifice and trouble, and they breed
More than they are bequeathed. So Beatrice,
Taking into her hands her garden tools—
A dream of a Campanian burial,
A trope of lightness, and a wild new world—
Begins the cultivation of the void.

References

Ananthaswamy, Anil. 2017. "Inside Knowledge: Is Information the Only Thing That Exists?" *New Scientist*, 29 March: https://www.newscientist.com/article/mg23431191-500-inside-knowledge-is-information-the-only-thing-that-exists/.

Arnold, Darrell. 2018. "Heraclitus, Fragment 94 (D 52)." *Heraclitus* blog, 29 June: https://philosophyandsociety.com/2018/06/29/heraclitus-fragment-94/.

Azarian, Bobby. 2022. *The Romance of Reality: How the Universe Organizes Itself to Create Life, Consciousness, and Cosmic Complexity*. BenBella Books.

"Computational complexity theory." 2025. *Wikipedia, The Free Encyclopedia*. Accessed February 27, 2025, https://en.wikipedia.org/wiki/Computational_complexity_theory.

Fraser, J. T. 1999. *Of Time, Conflict, and Human Values*. Illinois University Press.

Grundmann, Sven, Daniel Trabert, Kilian Fehre, Nico Strenger, Andreas Pier, Leon Kaiser, Max Kircher. 2020. "Zeptosecond Birth Time Delay in Molecular Photoionization." *Science* 370, no. 6514 (October 16): 339–41. https://doi.org/10.1126/science.abb9318.

Heller, Michael and Wiesław Sasin. 1998. "Emergence of time." *Physics Letters A* 250, no. 1 (December 21): 48–54. https://doi.org/10.1016/S0375-9601(98)00824-X.

Siegel, Ethan. 2024. "Did the Universe Have Zero Entropy When It First Began?" *Big Think*, February 22: https://bigthink.com/starts-with-a-bang/universe-zero-entropy/.

Sterne, Laurence. 2023. *The Life and Opinions of Tristram Shandy, Gentleman*. Penguin.

PART 2

Ourselves as the Measure of Time

CHAPTER 6

Pace and Time: Field Notes on the Art of Walking

Arkadiusz Misztal

Abstract

The paper examines the act of walking in relation to the measurement and experience of time. By considering ambulation as a basic spatial and temporal measure, the paper first discusses measurement techniques based on walking speed and pacing as intrinsic elements of what John Edward Huth has called "cultures of navigation." I argue that despite technological advances walking has retained an irreducibly qualitative character, which on the one hand make it impossible to produce a unified standard system of measurement and, on the other, reveal a peculiar dimension of embodiment and a lived sense of time. Through an active engagement of the body and the mind with the world, walking creates a sense of place and connection to the surrounding environment in its temporal dimension. The paper examines this engagement by drawing on the reflection and experience of two prominent walkers and thinkers: Henry David Thoreau and Maurice Merleau-Ponty. In so doing, it sketches a phenomenological account of walking as the art of navigating and exploring the plurality of temporalities underlying the manifold "scapes" of our lives.

Keywords

embodiment – motion – peripatetic – lived time – ambiguity – mensuration

1 Introduction: Mensuration and Pacing

> Above all, do not lose your desire to walk.
>
> SØREN KIERKEGAARD, letter to Jette (1847)

The term "pace" has accrued a number of meanings accommodating the wide range of contexts in which it has been used. The *Shorter Oxford English Dictionary* (2002) defines its primary meanings as a form of movement and a unit of length:

 | DOI:10.1163/9789004723634_008

> Pace: 1) a single separate movement made by the leg in walking, running, dancing, etc.; 2) the distance covered in a step by an individual; as a unit of length it refers to (i) the distance from where one foot is set down to where the other foot is set down (about 75 cm or 30 inches) (ii) the distance between successive stationary positions of the same foot (about 150 cm or 5 feet).
>
> *Shorter Oxford English Dictionary* (2002, 2063)

The second unit of length has been more common and more frequently applied in navigation and measuring techniques, in which "pace" refers to the distance covered by two steps of a walking person (left and right). Thus, for instance, a Roman pace, which is equal to 4.85 English feet, is measured from the heel of one foot to the heel of the same foot when it next touches the ground. It is this meaning of "pace" that I'll be referring to in the following section, which constitutes a backdrop for the phenomenological study of pedestrianism in the remainder of the chapter. In what follows I will explore pace as a form of movement and measure to highlight the variability of a unit that is typically imagined to be fixed and definite.[1]

The practice of counting paces has been employed for centuries in different cultures, as it is a natural measuring technique, which does not require sophisticated equipment. Pacing technique was widely used in ancient Greece and Rome, and provided a foundation for distance units such as the "league." In the Roman Empire 1000 paces equaled *mille passus* or *passuum*, now known as a Roman mile. The distances (within and between locations) were officially determined by professional pacers/surveyors trained to walk with equal-length steps. These pacers, called in Greek *bematists*,[2] [a *bema* in Greek is a step or pace, from *baino*, to walk, go] accompanied Alexander the Great on his campaign in Asia and were reported to have such precision that it has been suggested that they must have used an odometer (Engels 1980, 158).

To give another example, pacing technique was also employed by Eratosthenes in his most famous accomplishment: the measurement of the circumference of Earth. Using a well located in Syene (now Aswan, Egypt) as a reference point Eratosthenes noted that at noon on the summer solstice the

1 I would like to thank my anonymous reviewers for comments that helped improve the clarity of my argument.

2 Bematists (*bematistai*) were individuals who "measured distances by paces"; they are often mentioned together with *hemerodromoi* (runners of a day's distance), *dromokerykes* (running heralds), and perhaps chorographers. For a more detailed discussion of mensuration in ancient Greece, see Tzifopoulos (2012).

sun illuminated the entire bottom of the well without casting any shadows, indicating that the sun was directly overhead. He then measured the angle of a shadow cast by a stick at noon on the summer solstice in Alexandria and found it made an angle of 7.2 degrees. In order to finalize the experiment he used his knowledge of measurement (Eratosthenes was also a skilled *bematist*) to determine the distance between Alexandria and Syene and from there calculated the overall circumference of the earth.[3]

The practice of counting paces is still used today. Our smart watches, smartphones and fitness apps keep track of how many steps we make daily, weekly, monthly, etc. But pace count or the related "dead reckoning" technique[4] is also widely used in more professional contexts by Special Forces, infantry, SAR teams, forest service, hunters, and hikers. It proves particularly valuable in limited visibility conditions or as an integral element of micro-navigation when one needs a quick and reasonably accurate way of estimating distance.[5]

While pacing takes place primarily in regard to space, it is also naturally related to speed and time. As such it has been used as a measure for human-powered travelling. Although far from being universal, it has provided an uncomplicated standard shared among travelers. In ancient Persian and Arab cultures, a *parasang* or *farsakh* was a unit based on speed and time; the length of a *league*, derived from an ancient Celtic unit and adopted by the Romans as the *leuga*, is close to the *farsakh* and refers to the distance covered by a person in one hour. Correspondingly, the rate of three miles or five kilometers per hour, the speed at which most people walk on level ground (or how fast one can row a boat) has been used as a good approximation of human-powered travel. Nonetheless, human-powered travel is, as we know, inherently individual and responsive to the changes within and around the traveler. A healthy adult can walk as fast as six or seven kilometers per hour on smooth, level terrain. A walking pace is also dependent on a great number of factors, such as weather, type of terrain, altitude, types of shoes, companions, health and general well-being of the walker. Even the trained surveyor-bematists mentioned above could not always keep exactly the same pace. As a result, "units differed

3 Using his knowledge of geometry, Eratosthenes concluded that the shadow's angle would be the same as the angle measured from the Earth's center that cut off the arc between Alexandria and Syene. Assuming Earth to be a sphere, he used the principle that the center angle of a circle is proportional to the arc that that angle cuts off to calculate the overall circumference.

4 The method, originally used by sailors who would plot their course on a map, involves estimating one's position based on the direction and speed of travel from a known starting point.

5 For a lucid and comprehensive treatment of wayfinding strategies and techniques, see, for example, Walker (2016) and Caudill (2017).

from one kingdom to the next, causing confusion [as] Egyptian and Greek stadia differed" (Huth 2013, 55). The exact length of a stadium is unknown today, so that, in respect to Eratosthenes' experiment, "Values between 500 and about 600 feet have been suggested, putting the calculated circumference of the Earth between 24,000 and 29,000 miles" (Tretkoff 2006). A 5,000-mile discrepancy is hardly negligible, so that as technologically oriented modernity superseded the older paradigms the need for a standardization of measured units became both apparent and pressing. And yet these sorts of discrepancies are not merely historical anecdotes or records of crude, pre-scientific ways of experiencing the world in its complexity. Rather, they point to the intrinsically individual character of pacing, which cannot be simply reduced to a uniform rate of movement. They thus bring our attention to the peripatetic as a mode that is idiosyncratic and resistant to objective measurement.

Intrinsic to pacing is an individual gait that manifests itself through rhythm, tempo and a distinctive manner of walking. Pace cannot, therefore, be simply rendered as units of speed. For example, the speed of a person walking with short strides and a quick pace might equal someone walking with long strides and a slower pace. Even a basic fitness app or smartwatch aims to capture (at least to some extent) this individualized character of ambulation by recording the pace rate as the inverse of speed. If speed is measured in units of distance per unit of time (e.g. kilometers per hour), pace is measured in units of time per unit of distance (e.g. minutes per kilometer). This rhythmical dimension of walking is also evident even to an untrained observer, who can easily see, for example, how stride rhythm is coupled with breathing during locomotion. The particular pacing and its rhythm become even more apparent when, for instance, we start to use some sort of equipment such as walking poles, which change not only the synchronization of legs and arms but also affect the entire body posture and style of locomotion.

And there is more to walking than meets the eye. Walking does not merely set our bodies in motion, but also has remarkable powers of engaging body and mind with the surrounding environment. Anyone who has walked long enough knows that the seemingly simple act of ambulation can radically change the way our minds work and how we relate to the world, ourselves and others. In what follows I will explore the enigmatic reciprocity between a thinking mind and a walking body by drawing on the reflections and experience of two prominent walkers-cum-thinkers: Maurice Merleau-Ponty and Henry David Thoreau. Walking as a subject matter is wonderfully prolific and digressive; it offers many paths and avenues to explore. Walking, as Joseph A. Amato reminds us, "not only involves the story of feet and legs and the history of dress and footwear but also raises questions about roads, transportation, communication, cities, suburbs, and cultures. The act of going on foot is joined to a time, condition, society, and culture. Walking belongs to the gender, age, class, ethnic and national group, and even race of the walker" (Amato 2004, 2).[6] In what follows I will concentrate on walking in nature as conducive to a peculiar mode in which the walker can vividly and consciously experience a lived sense of place and time. The peripatetic, I argue, provides access to multiple temporalities and processes, which are often overlooked and marginalized in the technologically saturated environments we inhabit today.

2 The Lived Perspective of Walking and the Pace of Thoughts

Even seen as merely an act of biped locomotion, walking is not simple. It involves much more than keeping your balance and making steps. In order to move purposefully, we need to use our body to produce a trajectory, which requires coordination of four core elements of walking: posture, locomotor rhythm, gaze and gesture (Berthoz 2012, 120). These four elements are in turn dependent on the complex neural organizations and processes involved in creating a unified perception of the moving body in its relation to its environment. In our daily encounters with the world, we employ qualitatively different styles of ambulation (strolling, striding, trekking, hiking and so forth) to navigate the diverse "scapes" of our lives.[7] Inherent in all of these styles of locomotion and their corresponding trajectories is what Merleau-Ponty has called "the lived perspective" of the experiencing self.

The lived perspective reveals an embodied subject actively engaged in the specific context of the here and now. This embodied participation begins on a pre-reflective level as "the ongoing interchange between my [walking] body and the entities that surround it. It is a sort of silent conversation that I carry on with things, a continuous dialogue far below my verbal awareness" (Abrams 2017, 52). The relation between the lived body and the space in which it moves is neither neutral nor objective. As Merleau-Ponty puts it, we must

6 A peasant's shoes, to give an example, can reveal, as Martin Heidegger (1971) has shown, a rich and complex relationship between the wearer and the earth.

7 My use of this term owes much to Barbara Adam's work, especially her *Timescapes of Modernity*, where she studies "the complex temporalities of contextual being, becoming and dwelling" (Adam 1998, 11). Unlike other scapes such as landscapes, mountainscapes, or seascapes, which reveal the spatial qualities of past and present activities and the interactions of organisms and matter, "timescapes emphasise their rhythmicities, their timings and tempos, their change and contingencies" (11). Being shaped by the temporal dimension of living and its contextual practices, they are "the embodiment of practiced approaches to time" (11).

stop thinking of our body as merely placed "*in* space and *in* time. It inhabits [*habite*] space and time" (Merleau-Ponty 2012, 161, emphasis in original). The organic dynamics of the moving body reveals how the body occupies space and time "because movement is not limited to submitting passively to space and time, it actively assumes them, it takes them up in their basic significance" (Merleau-Ponty 2012, 117). In other words, space and time are not external to the lived body. Rather, they are experienced pre-reflectively as the phenomenal field in which my body perceives, acts and engages with the world. The temporal structure of the body is primarily directed towards the present moment, which retains the past: "At each successive instant of a movement, the preceding instant is not lost sight of. It is, as it were, dove-tailed into the present, and present perception generally speaking consists in drawing together, on the basis of one's present position, the succession of previous positions, which envelop each other" (Merleau-Ponty 2012, 162). The body also prepares for the future in anticipating impending positions, establishing "the link between a here and a yonder, a now and a future which the remainder of the instants will merely develop" (Merleau-Ponty 2012, 162). The body thus embraces the whole span of its movement by drawing together the past and the future. As Merleau-Ponty puts it, "In so far as I have a body through which I act in the world, space and time are not, for me, a collection of adjacent points nor are they a limitless number of relations synthesized by my consciousness, and into which it draws my body. I am not in space and time, nor do I conceive space and time; I belong to them, my body combines with them and includes them" (Merleau-Ponty 2012, 162). This sense of belonging reveals the body as open and indebted to the ambiguous modality of the world.

Walking also discloses this ambiguity by foregrounding motor intentionality as a source of practical knowledge, but of knowledge rooted in spatial and temporal openness to the world. The space in which I walk is not geometrical but practical and correlative to my living movement.[8] The walking body encounters its environment through motor intentionality and so invests it with meaning. My bodily awareness is both embodied and embedded within the environment. I encounter the world not as an abstract notion but rather experience it as intimately connected with my bodily subjectivity. This basic intentionality brings with it a modal awareness of belonging to the world.

8 In a similar vein, Husserl acknowledged the primacy of movement in the creation of meaningful experiences of the life-world. In his 1931 essay, "The World of the Living Present and the Constitution of the Surrounding World External to the Organism," Husserl argued that walking pre-reflectively draws on the concept of "I am in motion in space" and thus provides a sense of continuity between the "here" and the "theres" of my environment.

"Consciousness is in the first place not a matter of 'I think that' but of 'I can'" (Merleau-Ponty 2012, 159). I experience the environment in its specificity as the phenomenal field in which I can exercise my motor skills. The mountain stream, for instance, appears as passable, the rock as climbable if the skills required to negotiate them are part of a body schema that permits "a finely coordinated ensemble of motions intentionally organized in advance towards targets that are to be meaningfully moved" (Morris 2014, 252). My moving body is, as Edward S. Casey notes, at the same time kinetic and kinesthetic: moving and feeling-itself-moving (Casey 1993, 59). More precisely, I walk by virtue of combining the awareness of moving with a proprioceptive awareness of the body, which, apart from the kinesthetic sense of movement of my limbs and trunk, involves also a sense of balance, a sense of effort, the "internal" senses that provide information about nutrition and fatigue, and others.[9] My walking body thus incorporates lived attitudes to the world, which are practical and do not require conceptual representation.

Furthermore, the intentional motility released in walking reveals the ambiguous nature of bodily experience, which orients my perception and action towards the world of my concern. Merleau-Ponty distinguishes a whole spectrum of human ambiguities on both intentional and pre-reflective levels, "extending from the complex interrelationships between subjectivity and intersubjectivity, individuality and generality, to the dynamic interplay between my body and my historical situation, our intersubjective life and the world" (Weiss 2014, 288). Implicit in all of these kinds of ambiguities is a sense of the inconclusiveness of temporality as it is lived: "My hold on the past and the future is precarious and my possession of my own time is always postponed until a stage when I may fully understand it, yet this stage can never be reached, since it would be one more moment bounded by the horizon of its future, and requiring in its turn, further developments in order to be understood" (Merleau-Ponty 2012, 346). As a result, temporal synthesis is always incomplete, as Weiss declares in pointing out that, "I'm not contained in the 'now' of the present" (Weiss 2014, 292). Rather, temporal experience extends organically to the past and the future, as Merleau-Ponty describes: "my living present opens upon a past which nevertheless I am no longer living through, and a future which I do not yet live" (Merleau-Ponty 2012, 433).

By confronting the inherent nature of temporal ambiguity I experience both the ambiguity of being in the world and the body as a locus of self-transcending. The inherent inconclusiveness of my existence and its directedness towards

9 For a lucid overview of the history and problematics of proprioception, see Uwe Proske, Simon C. Gandevia (2012).

the future preclude my ever being self-present and inscribed into the web of causal patterns and structures. This inconclusiveness manifests itself in the idiosyncratic dynamics of motility. The moving body is, as Casey points out, inherently ambiguous, specifically in its relation to measure:

> [the body's] elliptical saunterings and sidewise turnings (its literal con-siderations), *defy calibrated measurement*. Even its straight courses are never rectilinear; even when I am moving unequivocally toward a given goal, I continue to turn and twist my serpentine body along the way. The amorphousness of my lived body [...] ensures that my connection with the near and the far will be indeterminate, both in space and time.
>
> CASEY 1993, 59; emphasis added

When moving across a treacherous scree field or when wading a stream, my entire body continually adapts to the changing conditions (slippery stones, steepness of slope, wind direction, water temperature, etc.) without the help of my verbal consciousness, which, when activated, tends to interfere with and obstruct the movement rather than help. In this lived engagement with the surrounding environment the embodied self actively participates with the sensuous terrain. The alpine scree field or the strong current of a narrow mountain stream are not features that I merely reflect on. They are experienced as elements of a living landscape in which I am corporally embodied: they disclose "the structure of the landscape as an emerging organism" (Merleau-Ponty 2004, 281) as it is lived from within the living, attentive body.

The embodied participation inherent in walking thus binds us integrally with the landscape, and does so by opening us up to processes and activities that might otherwise be seen as mere measuring. Walking, as Philip Gros has argued in his *Philosophy of Walking*, "causes absorption" (2014, 85) that taps into a close and intimate continuity between the walking body and the body of the land. When hiking at length in the mountains, "[t]he body becomes steeped in the earth it treads. And thus, gradually, it stops being in the landscape: it *becomes* the landscape" (85, original emphasis). This participation "structures a perceptually sensitized, kinesthetic exploration of both macro and micro features of the landscape" (Tafler 2019, 55), and so radically changes our perspective: we do not merely move through or across the landscape; we walk as a part of it. The "vibrations" of the landscape[10] are repeated in the walker's

10 These vibrations are not to be thought of as merely physically acoustic stimuli; rather, in a more general sense, they render the absolute fullness of the living landscape as it is given

body (Gros 2014, 24). Inner and outer are synchronized as the rhythm of the moving body is aligned with the rhythm of the landscape. This is perhaps most clearly detectable in the measure of our changing breath and heart rates[11] as they adjust to the texture of the terrain, changes of altitude, the encountered weather situation, etc.

Walking, however, is also "an ambulation of mind," to use Gretel Ehrlich's phrase (1990, 28), or, as Rebeca Solnit suggests in *Wanderlust*, "the rhythm of walking generates a kind of rhythm of thinking" (2000, 5). This creative peripatetic power invigorated the thoughts not only of ancient Greek philosophers but also of modern walkers-cum-thinkers such as Henry David Thoreau, thinkers whom Nietzsche explicitly encouraged to "think while walking, walk while thinking, and let writing be but the light pause, as the body on a walk rests in contemplation of wide open spaces" (qtd. in Gros 2014, 20). The embodied self engaged in walking is bound to encounter not only open space but also that familiar stranger, Time, to use J. T. Fraser's expression. By combining the unconscious activity of the body with the conscious activity of the mind, walking, as Solnit writes, "hits a steady beat that seems to be the pulse of time itself" (30). In this carnal-noetic alignment of measure, "thinking becomes almost a physical, rhythmic act" (Solnit 2000, xv), capable of traversing not only space but also time.

3 Walking with Time

Walking offers a kind of release from temporal routines and time measures, from the monotony of daily life and its "time squeeze." This temporary disconnection from time regimes can be experienced, for example, during long hikes or multi-day treks when one no longer knows how many hours, or even days

---

in the experience of walking. As such they are akin to the "vibration of appearances" that Cezanne sought to capture in his paintings (Merleau-Ponty 2004, 281).

11 Neurobiological research confirms that walking benefits brain function and mental health: a rhythmical form of synchronization has been recorded in a recent study that reveals a close correlation between ambulating and cerebral blood flow. Researchers at New Mexico Highlands University (NMHU) found that walking can positively affect the supply of blood to the brain, specifically as it matches our own biological measures. "'What is surprising is that it took so long for us to finally measure these obvious hydraulic effects on cerebral blood flow,'" first author Ernest Greene explained. "There is an optimizing rhythm between brain blood flow and ambulating. Stride rates and their foot impacts are within the range of our normal heart rates (about 120/minute) when we are briskly moving along" (*Experimental Biology* 2017).

have passed. The release does not happen automatically, but must be willed. As Thoreau, a seasoned walker, confesses "[o]f course it is of no use to direct our steps to the woods, if they do not carry us thither. I am alarmed when it happens that I have walked a mile into the woods bodily without getting there in spirit" (Thoreau 2001, 229). The carnal-noetic alignment is gradual, as it requires a smooth integration of walking's outward context with its inward density, as seen in "the accompanying mental, visual, and perceptual processes that [...] interconnect the individual with the surrounding environment" (Forgione 2005, 688). This synchronizing adjustment is achieved by "finding your pace," by adjusting breathing rhythm and movement of the legs so that, as Gros puts it, "they do not shatter but [...] *make the world's presence palpable and keep time with it*" (2014, 212, original emphasis). One can then connect with the measured movement of the sun and stars on the horizon, with cycles of mornings and evenings, and "echo with the slow tread the cadence of the minutes, hours and days" (Gros 2014, 78). This art of walking does not kill time but seeks "to welcome it, to pick off its leaves and petals one by one, second by second" (Gros 2014, 190). One learns thus to walk with time, "matching its pace as one does with a child" (Gros 2014, 78). This sort of slowness resonates strongly with Solnit, who admits in *Wanderlust*, a book conceived as a result of walking: "I like walking because it is slow, and I suspect that the mind, like the foot, works about three miles per hour" (10).

Thoreau tested this proposition in his intellectual, personal, and professional pursuits. The peripatetic permeated not only his writings and lectures but also his scientific and professional activities. He was also a measurer by nature, and combined his almost compulsive habit of taking long walks with professional occupation by working as a surveyor in the Concord area. His journal, the daily practice of writing that accompanied his walks, reflects his keen interest in methods of scientific observation and the exact calculation of distance, area, volume, and time. The journal comprises notes and lists of measurements as well as meticulous descriptions of natural phenomena. One of Thoreau's projects, for example, consisted in assembling an indexed chart, which he called a "kalender," charting the seasonal life of Concord in a way that recorded temporal cyclicity while retaining a lived sense of the passing moment. This strong interest in ways of measuring time surfaces also in *Walden*, where Thoreau "proposes to erect in the stream of time a 'Realometer' rather than an Egyptian 'Nilometer' to measure the significance of the transcendental moments" (Guthrie 2001, 131).

Alongside his interest in measurement, Thoreau was acutely aware of the tension between time determined by economic, cultural and social constraints and time experienced naturally, time in the wild. Walking not only resolved this tension by permitting him to escape "meridian (i.e. noontime) social dinners that threaten to waste his time" (Guthrie 2001, 149) but also sharpened his receptiveness to many forms of temporality that are normally overlooked. Accordingly, his essay. "Walking,"—his most explicit treatment of the art—is animated by the imperative to live deliberately in the moment, to expose oneself "to a wide spectrum of possibilities radically different from habits, routines, and patterns of everyday life" (Misztal 2025). The idea of growing in the present is central to Thoreau's perspective on time and is most strongly articulated in his image of corn growing in the night: "I believe in the forest, and the meadow, and the night in which the corn grows" (Thoreau 2001, 240). This growth requires an interval of 'no-time' during which momentous processes unfold. It can be achieved only if "we can learn to live extemporaneously, as part of nature" (Guthrie 2001, 159). As Guthrie helpfully notes, Thoreau used the image of corn's nocturnal growth as part of a description of night-walking, presented in his journal as an act of exploring alternative reality. "Many men walk by day; few walk by night. It is a very different season" (Thoreau 2009, 92). Night is the period of different temporality, of "no-time" in which "growth is accomplished not through activity and conscious effort, but through passivity, relaxation, and unconscious effort" (Guthrie 2001, 158). By exposing oneself to periods of no-time, that is non-measurable temporalities, "one learns to grow vertically and thus provides depth to the present moment" (158).

Another potent metaphor for an alternative temporal modality is that of the Old Marlborough Road, invoked by Thoreau in the poem incorporated into "Walking." First traced in the 17th century, the road functioned as a commercial route connecting the town of Marlborough with Concord. By Thoreau's time the road had fallen into disuse and no longer had any economic value.[12] Traveling along the abandoned route allowed Thoreau to immerse himself in purely "idle" time unconcerned with business. As Thoreau wrote, "With the instinct to travel, / I can get enough travel / On the Old Marlborough Road. / Nobody repairs it, / For nobody wears it" (Thoreau 2001, 232). Strolling alone and without purpose, Thoreau not only practiced the Nietzschean imperative to "think while walking" but also sought transcendental moments, only experienceable outside of conventional time. Walking a discontinued road that led to nowhere, Thoreau was simultaneously moving outward and inward and thus could enter "the liminal, timeless state of mind that he called a 'revery' and treated as a physical location in its own right" (Guthrie 2001, 132). The art of sauntering was a mode of both walking with and out of time.

---

12 For a more detailed discussion of the Concord area in connection with Thoreau, see Gross (2022).

In *Walden* Thoreau makes it clear that he prefers a walking stick to a surveyor's leveling rod, notching the present moment on his stick, much like the artist of Kouroo cited in the concluding chapter of the book. In the Hindu parable the artist resolves to create a perfect work, and so devotes his entire life to the slow pursuit of perfection: "One day it came into his mind to make a staff. Having considered that in an imperfect work time is an ingredient, but into a perfect work time does not enter, he said to himself. It shall be perfect in all respects, though I should do nothing else in my life" (Thoreau 1985, 243). Fully dedicated to the task, the artist gradually attains his goal by unwaveringly following his belief and making "no compromise with Time" (243). True to his purpose, he is limited neither by the passing of the dynasty of the Candahars nor by the transition of the star Kalpa from the earth's northern pole.[13] In making the staff he creates "a new system" in which time is not an obstacle. He comes thus to realize that "the former lapse of time had been an illusion, and that no more time had elapsed than is required for a single scintillation from the brain of Brahma to fall on and inflame the tinder of a mortal brain" (243). As Guthrie points out, Thoreau skillfully incorporates into this parable the themes of temporal mensuration and creative resolution:

> Although the artist of Kouroo does not realize it, he is not just a carver, but also a measurer, like Thoreau; more than that, he is himself a unit of measure, for as soon as he has finished work on his stick, a cycle of time much larger than a kalpa culminates, and a new order begins. Thus it is the carver, and not the stick itself, that determines the 'length' of time.
>
> GUTHRIE 2001, 243

Thoreau likewise seeks to determine time by means of creative endeavors and artistic pursuits. His art of sauntering involves an engagement with the environment in its historical specificity and locality. Walking, as Solnit notes, "is how the body measures itself against the earth" (Solnit 2000, 31), and yet this measure resists quantification and commodification. A case in point is the oft-quoted paradox of travelling to Fitchburg. As Thoreau points out, one

13 As Guthrie points out, Thoreau mistakenly identifies "kalpa" as the Hindu word for the polar star. In contrast Thoreau remarks in a journal entry from 1842 that kalpa refers to "a unit of time, defined as one day in the life of Brahma, or 320,000,000 years" (Guthrie 2001, 170). And yet in the above quoted passage, Thoreau retains the meaning of "kalpa" as denoting vast stretches of time by invoking the image of Brahma awaking and slumbering many times.

actually covered the thirty mile distance between Concord and Fitchburg more swiftly by walking than by taking a train, which required earning and paying the 90 cents fare, approximately a day's wage:

> the swiftest traveller is he that goes a-foot. […] Well, I start now on foot, and get there before night […]. You will in the meanwhile have earned your fare, and arrive there some time to-morrow, or possibly this evening, if you are lucky enough to get a job in season. Instead of going to Fitchburg, you will be working here the greater part of the day.
>
> THOREAU 1985, 41

More importantly, however, walking, as artful and purposeful exertion, elevates and refines our engagement with the world. It provides access to the primordial vitality of the living landscape that cannot be categorized or rationalized. Sauntering, for Thoreau, as Susan McWilliams observes, opens up the present so that within a single moment immense stretches of time can be experienced; sauntering can give you "experience of an interval as between one life and another—a greater space than you ever travelled" (Thoreau 2009, 218).

Not unlike the artist of Kouroo, Thoreau dedicated his life to carving and shaping his time, to notching the present on his walking stick: "In any weather, at any hour of the day or night, I have been anxious to improve the nick of time, and notch it on my stick too; to stand on the meeting of two eternities, the past and the future, which is precisely the present moment" (Thoreau 1985, 16). Thoreau does not intend to survey the line between the two eternities, does not aim to measure it, not even by the pacing technique mentioned above, but instead seeks to walk it, "to toe that line" (Thoreau 1985, 16) so that he can amplify the moment and access its spectrum of possibilities. Thoreau, like Kierkegaard, never lost the desire to walk. By practicing the art of sauntering he walked himself into his best thoughts and showed that we "gain nothing by transferring into ourselves the time that belongs to things" (Merleau-Ponty 2012, 479). The embodied self of the walker neither constructs time nor finds it in the world, but meets it as its own dimension of being. As Merleau-Ponty puts it, "I am not the creator of time any more than of my heart-beats. I am not the initiator of the process of temporalization; I did not choose to come into the world, yet once I am born, time flows through me, whatever I do" (Merleau-Ponty 2012, 496). By keeping pace with time, one can tap into this flow, into this "ceaseless welling up" (Merleau-Ponty 2012, 497) that underlies our embodied existence and shapes the time that one lives into "a new and original system."

References

Abrams, David. 2017. *The Spell of the Sensuous: Perception and Language in a More-Than-human World.* New York: Penguin.

Adam, Barbara. 1998. *Timescapes of Modernity: the Environment and Invisible Hazards.* London: Routledge.

Amato, Joseph. 2004. *On Foot: a History of Walking.* New York: New York University Press.

Berthoz, Alain. 2012. *Simplexity: Simplifying Principles for a Complex World.* New Haven, CT: Yale University Press.

Casey, Edward S. 1993. *Getting Back into Place: Toward a Renewed Understanding of the Place-World.* Bloomington: Indiana University Press.

Caudill, Craig. 2017. *Extreme Wilderness Survival.* Salem: Page Street Publishing.

Ehrlich, Gretel. 1991. *Islands, the Universe, Home.* New York: Viking.

Engels, Donald W. 1980. *Alexander the Great and the Logistics of the Macedonian Army.* Berkeley: University of California Press.

Experimental Biology. 2017. "How Walking Benefits the Brain." Accessed January 31, 2023, https://neurosciencenews.com/neurobiology-walking-6487/.

Forgione, Nancy. 2005. "Everyday Life in Motion: The Art of Walking in Late-Nineteenth-Century Paris." *The Art Bulletin* 87 (4): 664–87.

Fraser, J. T. 1987. *Time: the Familiar Stranger.* Amherst: University of Massachusetts Press.

Gros, Frédéric. 2014. *A Philosophy of Walking.* London: Verso.

Gross, Robert A. 2022. *The Transcendentalists and Their World.* New York: St Martin's Press.

Guthrie, James R. 2001. *Above Time: Emerson's and Thoreau's Temporal Revolutions.* Columbia: University of Missouri Press.

Heidegger, Martin. 1971. "The Origin of the Work of Art." In *Poetry, Language, Thought.* New York: Harper & Row: 15–88.

Husserl, Edmund. 1981. "The World of the Living Present and the Constitution of the Surrounding World External to the Organism." In *Husserl: Shorter Works,* edited Peter McCormick & Frederick A. Elliston. Notre Dame: University of Notre Dame Press.

Huth, John. 2013. *The Lost Art of Finding Our Way.* Cambridge: The Belknap Press of Harvard University Press.

McWillliams, Susan. 2009. "Thoreau on Body and Soul." In *A Political Companion to Henry David Thoreau,* edited by Jack Turner. Lexington: University Press of Kentucky.

Merleau-Ponty, Maurice. (1964) 2004. "Cezanne's Doubt." In *Maurice Merleau-Ponty: Basic Writings,* edited by Thomas Baldwin. London: Routledge Classics.

Merleau-Ponty, Maurice. (1962) 2012. *Phenomenology of Perception*. Translated by Colin Smith. London: Routledge & Kegan Paul.

Merleau-Ponty, Maurice. (1964) 1968. *The Visible and the Invisible*. Translated by Alphonso Lingis. Evanston: Northwestern University Press.

Misztal, Arkadiusz. 2025. "As Wild as Possible: Modal Sauntering in Henry David Thoreau's Thought." Manuscript in preparation.

Morris, David. 2014. "Body." In *Merleau-Ponty Key concepts*, edited by Rosalyn Diprose and Jack Reynolds. London: Routledge.

Nietzche, Fredrich William. 1911. *The Twilight of the Idols*. Volume 16, translated by Anthony M. Ludovici. Edinburgh: T. N. Foulis. Accessed February 25, 2023, https://babel.hathitrust.org/cgi/pt?id=mdp.39015005117109.

Proske, Uwe and Simon C. Gandevia. 2012. "The Proprioceptive Senses: Their Roles in Signaling Body Shape, Body Position and Movement, and Muscle Force." *Physiological Reviews* 92 (4): 1651–97.

Shorter Oxford Dictionary on historical principles. 2002, 5th edition, New York: Oxford University Press.

Solnit, Rebecca. 2000. *Wanderlust: A History of Walking*. New York: Viking.

Tafler, David. 2019. "*World-Wide-Walks*: body apparatus." In *World-Wide-walks: Crossing Natural-Cultural-virtual-frontiers*, edited by Peter d'Agostino and David Tafler. Bristol, UK: Intellect.

Thoreau, Henry David. 1985. *A Week On The Concord And Merrimack Rivers, Walden; or, Life In The Woods, The Maine Woods, Cape Cod*. Edited by Robert F. Sayre. New York: Library of America.

Thoreau, Henry David. 2001. "Walking." In *Collected Essays and Poems*, 225–55. New York: Library of America.

Thoreau, Henry David. 2009. *The Journal of Henry David Thoreau, 1837–1861*. New York: National Geographic Books.

Tretkoff, Ernie. 2006. "Eratosthenes Measures the Earth." *APS Journal* 15, no. 16. Accessed January 15, 2024, https://www.aps.org/publications/apsnews/200606/history.cfm.

Tzifopoulos, Yannis. 2012. "Bematists the Encyclopedia of Ancient History." John Wiley & Sons. Accessed January 25, 2024, 10.1002/9781444338386.wbeah09247.

Walker, Brian. 2009. "Thoreau's Alternative Economics: Work, Liberty, and Democratic Cultivation." In *A Political Companion to Henry David Thoreau*, edited by Jack Turner. Lexington: University Press of Kentucky.

Walker, Kevin. 2016. *Mountain and Moorland Navigation*. Gwyneed: Pesda Press.

Weiss, Gail. 2014. "Ambiguity." In *Merleau-Ponty Key concepts*, edited by Rosalyn Diprose and Jack Reynolds. London: Routledge.

CHAPTER 7

Taking Chronobiology Seriously

Kerstin Cuhls, Friederike Mork-Antony, Ralph Gutknecht and Thomas Kantermann

Abstract

The German chronobiology project CIRCADIA started in 2021 to investigate how circadian rhythms are influenced by new and variously combinable technologies in everyday life, an issue which became more important during the Covid-19 pandemic. The constant de-structuring of everyday life that such technology can influence, with the consequence of changes in sleep time and a de-rhythmisation of biology and psychology is particularly dangerous, as many people are unaware of structures in time, timing and their biological rhythms. Through a systematic stocktaking and foresight (literature analysis, horizon scanning, online survey, explorative workshops, impact assessment) of changes due to the pandemic, CIRCADIA took a short- (5 years) and a long-term look (10 and more years) at the potential effects of an increased use of digital devices. On the basis of our results, we will develop both prevention and solution strategies for existing problems. This paper accordingly raises the question of measuring time in the rhythms of daily life and presents results of a representative survey that measured these rhythms and matched them to chronotypes and changes in the use of technology since 2019.

Keywords

chronobiology – circadian rhythms – foresight – horizon scanning – chronotype

Humans have innate circadian rhythms that are inextricably linked to the regulation of their sleep-wake timing and body's homeostasis. Stable circadian rhythmicity depends on undisturbed exposure to distinct *Zeitgebers*, or time-cues, particularly exposure to light. Disruptions of circadian rhythms are associated with sleep disturbances, cognitive impairment, and health problems.

 | DOI:10.1163/9789004723634_009

In this chapter, we introduce a project called CIRCADIA. We started this project with an inventory of researched topics in the realm of human chronobiology and added a new outlook by raising questions concerning future developments in the field and its context via semi-automated horizon scanning. In 2022, we conducted a representative survey in Germany which addressed the COVID-19 pandemic and asked about daily behaviors, including sleep and work-related aspects, as well as daily use of technology. We then invited experts from distinct research areas and companies to three workshops to both verify the findings and help design recommendations on how to make use of this knowledge. The results from the workshops and the recommendations are reported in three Policy Briefs[1] and a recent paper (Cuhls et al. 2024).

1 What Are Circadian Rhythms?

The human body is equipped with a circadian system that helps control homeostatic processes and coordinate physiology, behavior, sleep and wakefulness across the 24-hour day (Hastings, Maywood, and Reddy 2008). As its name implies (Latin *circa* = around, *dies* = day, so "around a day"), the period length of the circadian rhythm in most people is not exactly 24 hours, but rather 24.2 hours on average (Czeisler et al. 1999). For the circadian system to be entrained to the 24-hour day, it must be adjusted daily. Environmental signals able to make this adjustment are called *Zeitgebers*, of which light is the strongest for the human circadian system, mediated via the suprachiasmatic nucleus (SCN) of the anterior hypothalamus (Hattar et al. 2002; Münch et al. 2020).

Circadian rhythms with a period length longer than 24 hours need to be slowed down, because they tend towards later entrainment. Circadian rhythms with a period length shorter than 24 hours need to be accelerated, because they tend towards earlier entrainment. Light information processed via the SCN does not convey information to the visual system, but rather helps generate non-visual effects of light which support circadian entrainment. These non-visual effects of light depend on light intensity, spectrum, duration of exposure, time-of-day, spatial distribution, and light history (the amount of light the circadian system was exposed to in the past hours) (see Figure 7.1). As a result, the impact of light on the circadian system varies with time of day

1 They can be found on the website: https://www.isi.fraunhofer.de/de/competence-center/foresight/projekte/circadia.html#1.

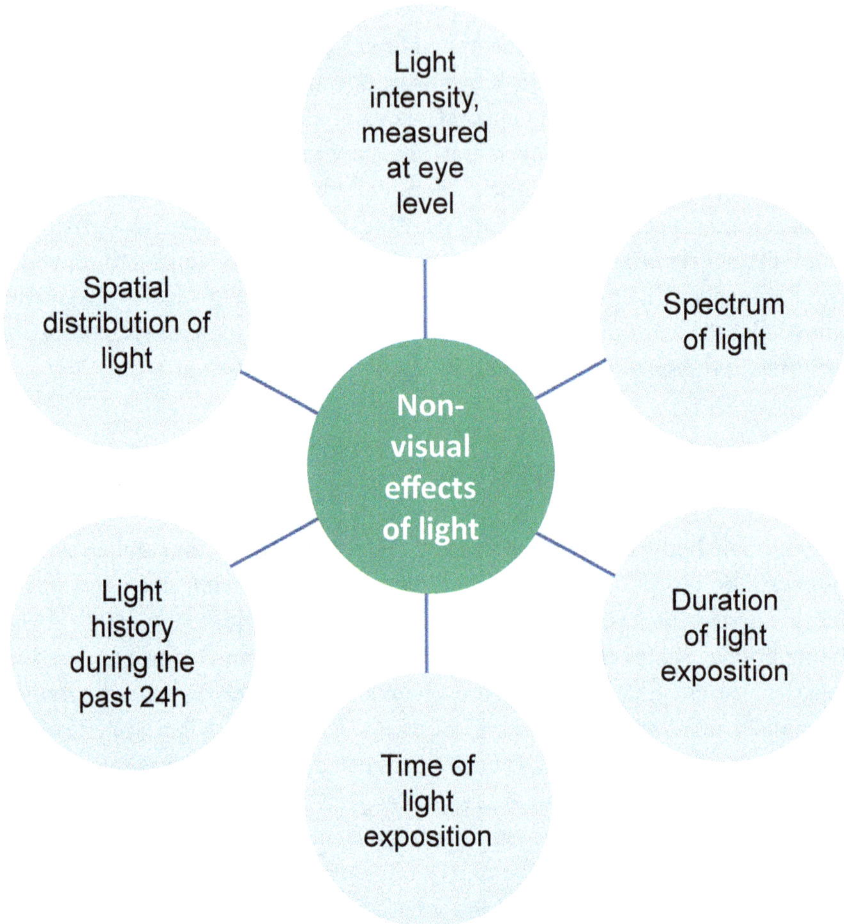

FIGURE 7.1 Factors related to the non-visual effects of light on the human circadian system
SOURCE: GERMAN STATUTORY ACCIDENT INSURANCE 2018—DGUV INFORMATION 215–220

and intensity. Light during the subjective morning (based on internal time) can advance the phase of the circadian system and the timing of sleep. Light during the subjective evening and night (before the core body temperature reaches its minimum) can delay the phase of circadian system and sleep timing (Roenneberg et al. 2007; Minors, Waterhouse, and Wirz-Justice 1991). Because humans differ in the period length of their circadian rhythm and in the patterns of their daily exposure to light, they also differ in chronotype, from early "larks" to late "owls" (Roenneberg et al. 2007; Roenneberg, Anna Wirz-Justice, and Merrow 2003).

2 Disruptions to Health and Well-Being

Challenges to the human circadian system are well described in the literature, ranging from minor challenges like daylight saving time (Kantermann et al. 2007; Roenneberg, Winnebeck, and Klerman 2019) and early school hours (van der Vinne et al. 2015; Zerbini et al. 2017) to major challenges like shift- and night work (Arendt 2010; Costa 2003; Hansen and Lassen 2012; Kantermann et al. 2012; Ritonja et al. 2019). These studies document, for example, psychological disturbances, performance deficiencies, metabolic disorders, and cardiovascular problems. Our study, moreover, adds pain, migraine, ADS/ADHS, and diabetes to that list (Cuhls et al. 2024). At the level of the individual, there is evidence that other behaviors, such as screen use in the evening, can also have a negative impact on sleep (Basner and Dinges 2009; Basner, Spaeth, and Dinges 2014). Such studies focus on the light emitted from screens and its potential impact on the circadian system (Cajochen et al. 2011; Hale et al. 2018). These health and performance problems constitute a societal strain. In 2017, Hafner and his colleagues calculated that the economic costs of insufficient sleep (for example, in reduced workplace productivity) came to very high sums, such as 60 billion dollars per year in Germany, 411 billion in the USA and 138 billion in Japan. These figures amount to 1.56% of the GDP of Germany, 2.28% in the USA and 2.92% of the Japanese GDP (Hafner et al. 2017).

3 The Project CIRCADIA—Circadian Rhythms and Technology

In November 2022, we conducted a representative survey among 2,000 people aged 18 and older in Germany. Data collection took part within the project CIRCADIA, which was a project funded by the Federal Ministry of Education and Research (BMBF) as part of the INSIGHT program on interdisciplinary perspectives on social and technological change (Project ID: 16INS106B). Participants were asked to fill in questions about their everyday life, work, sleep, and their use of various forms of technology, particularly in the context of the COVID-19 pandemic. Our aim was to examine the impact of technology use on sleep and daily rhythms. The results were used to help develop coping strategies to assist a better alignment of an individual's chronobiology and their tasks in daily life.

The project focused on technology use and its changes during and after the pandemic. We concentrated on forms of technology

a) … which are light-emitting and have a direct effect on the circadian system (e.g. LED, smartphone display).
b) … that contribute to the de-structuring and de-rhythmization of everyday life and so can indirectly influence the circadian system of users (e.g. online collaboration tools, internet, social media applications).
c) … which can contribute to adapting daily routines to circadian rhythms (e.g. self-measurement apps, web-based coaching, smartphone-based assistance systems).

The project started with a literature search for the years 2017–2022 via Pubmed, Google Scholar and other databases. We analyzed the literature in the triangle of technology use, social practices, and chronobiology/circadian rhythm research. Key aspects examined in the literature search include:

- Daylight exposure: depends on e.g. weather, health status, age, socioeconomic status, immigration status, day of the week;
- Daily routines: related to improved mood, sleep and circadian rhythm entrainment;
- Bedtimes/sleep conditions: impact on the quality and quantity of sleep;
- Daytime napping: can affect melatonin secretion in duration and timing;
- Personal hygiene: e.g. warm showers/baths can have a positive effect on sleep;
- Working and school times/working conditions: key factors include shift- and night work, light exposure at work, architecture (e.g. indoor lighting, number of windows), screen time at work;
- Hobbies/leisure activities: e.g. training schedules by chronotype, outdoor/indoor training;
- Stress, personality traits, attitudes towards sleep: the impact of conscientiousness and sleep-related beliefs on sleeping habits;

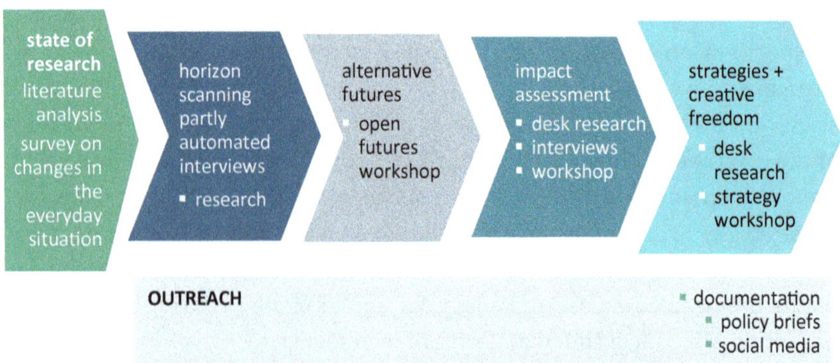

FIGURE 7.2 Project CIRCADIA steps
SOURCE: FRAUNHOFER ISI/FOM

- Living conditions: the influence of parents' (particularly mothers') and children's sleep on each other;
- Urban planning and architecture: light at night (ALAN), lack of daylight.

The second step in the project was a semi-automated horizon scanning concerning potential future developments, using the database Dimensions.ai. The horizon scanning worked well for the scientific findings, but less well for the societal developments. For societal developments, we did a manual screening of journals. In the horizon scanning and in a 'futures workshop' that included experts from chronobiology and other research areas, persons from companies, safety personnel, teachers, entrepreneurs, and others, we identified drivers and topic-clusters that may affect circadian rhythms and sleep. These drivers include demographic and climate change (e.g. concerning heat at night), societal aspects (i.e. dilemmas between employment duties, preparing kids for school, and personal circadian rhythms), challenges to sleep such as daylight saving time and also personal behavior. We grouped the drivers into four topics:

1. Sleep,
2. Bounded compared to unbounded time,
3. Lighting technology, and
4. Safety and security.

4 CIRCADIA Survey—First Results

4.1 *Demographics, Sleep Timing and Quality*

Altogether, 1922 data sets (976 women, 935 men, 11 non-binary persons, aged 18–85 years) were suitable for analysis. Of the respondents, 59.3% were employed, 3.6% were unemployed (job seekers) and 36.1% were not working (e.g. retired, pupils, students without part-time employment). Most participants were employed full-time (34.6%) or retired without secondary employment (27.7%). Of the remainder, 246 people were working part-time (12.7%).

Participants took longer to fall asleep on workdays (on average approximately 20 minutes) than on work-free days (on average approximately 18 minutes). Most respondents rated the quality of their sleep on workdays as "good" (39.9%) or "neither good nor bad" (32.6%), while 12.5% stated that their sleep quality was "poor." For 3.9%, their sleep quality was "very poor" and for 9.8% "very good." Overall, the quality of sleep on workdays was rated worse than on work-free days.

On work-free days, most people rated the quality of sleep as "good" (47.6%) or as "neither good nor bad" (26.9%). "Very good" was given by 14.9%, "poor" by 7.5% and "very poor" by 2.6% of respondents. The reasons given by the participants for "poor" or "very poor" sleep quality were that "they woke up on their own" during the night (54.9% and 58.2% respectively) or "had difficulty falling asleep" (38.2% and 38.6% respectively).

Many participants stated that on workdays they "suffered from tiredness the next day" (38.6%), while the figure was lower on work-free days (28.4%). The average duration of sleep reported by all participants was shorter on workdays (7 hours, 22 minutes) than on work-free days (8 hours, 11 minutes). Most people did not use sleep aids (61.4%). However, if aids were used, these were most frequently the television (18.0%) or technical devices without a screen (11.2%), followed by sleeping pills and other medication (7.3%) and melatonin supplements (5.8%).

In the subgroup of participants stating that they had experienced changes in sleep since the Covid-19 pandemic, 38.8% indicated that they needed more time to fall asleep and 44.3% stated that they woke up more often at night, at least some of the time. For around 10% on workdays and 8% on work-free days, it took longer than 30 minutes to fall asleep. Reasons given for taking longer than 30 minutes to fall asleep were "I can't switch off, my mind is racing" (75.3% on workdays and 66.9% on work-free days), "I'm restless or nervous" (33.6% and 31.1%, respectively) and "I'm not tired" (17.5% and 24.0%, respectively).

4.2 *Technology Use and Bedtime Procrastination*

One section in the survey focused on the use of screen media during the day and in the last hour before going to bed, distinguishing between workdays and work-free days. We found that on workdays 14.4% of respondents did not use screen media before going to bed, while on work-free days the number was 12.7%. Across all participants, most people used the TV before going to bed both on workdays (56.1%) and on work-free days (61.6%), followed by their smartphone (43.2% and 44.8%, respectively) and laptop/computer (17.5% and 20.3%, respectively). The e-book/tablet was mentioned by 10.9% on workdays and 12.7% on work-free days and a games console by 3.4% on workdays and 5.0% on work-free days. Older people were more likely not to use any device, and also more likely to use the TV, compared to younger people. Younger people, in turn, were more likely to use a smartphone, laptop, or games console.

When participants were asked what they usually do when they wake up at night, only a few stated that they used screen media. A proportion of 6.3% (workdays) and 8.7% (work-free days) turned on the TV or used their smartphone (4.0% and 5.5%, respectively). Slightly more participants used screen media on work-free days, compared to workdays, when they woke up at night. Participants most frequently reported themselves as staying in bed and trying to go back to sleep (45.6% and 48.0%, respectively).

With respect to changes in technology use since the start of the Covid-19 pandemic, we found that more than half of respondents (52.6%) have at least partially increased the amount of time spent with friends or have increasingly met others via a screen. In terms of total screen time, half (50.4%) of participants reported an increase. The time spent on video conferences increased for 44.5% of participants, while webinars increased for 41.6%. Spending more time playing video games was reported by 34.6%. More than 20% of respondents used devices with screens as a sleep aid. More than half of respondents used them daily.

Participants were asked the following question to find out about bedtime procrastination: "Sometimes people end up staying awake longer than they originally wanted, even though they know it's not necessarily good for them (e.g. because they're not getting enough sleep). What is it normally like for you?"

In response to this question 57.9% answered that they had not stayed up longer than planned in the last four weeks. Around 39% of respondents stated that they usually postponed going to bed (both on workdays and work-free days). Reasons given for the delayed bedtime were to finish watching TV (20.3% on workdays and 23.7% on work-free days), not being tired (14.0% and 12.0%, respectively), surfing the internet (13.8% and 15.5%, respectively) or being unable to switch off due to stress and worries (13.6% and 10.7%, respectively). For some, socializing with friends or acquaintances (10.4% and 12.1%, respectively) led to a delay in going to bed. The fear of missing something on social media was rarely mentioned (3.7% and 3.8%, respectively).

On workdays, participants more frequently cited worries and stress as a reason for postponing bedtime than on work-free days. Postponing bedtime due to the use of screen media, on the other hand, was mentioned more frequently on workdays than on work-free days.

In both cases (Figures 7.3 and 7.4), the smartphone has almost caught up with the TV in terms of frequency of use in the hour before falling asleep (both devices can be used at the same time). Slightly more people reported using a laptop or computer than reported using "no device" before sleep. Screen media were used more commonly on work-free days than on workdays. When we discussed these results in our 2nd project workshop, participants anecdotally reported that on work-free days, they use screen media more often because that is the time they can relax (me-time). It also was stated that the evening is the only time of the day they have for themselves, when their children are in bed and work is done, so that they can use this time for their own purposes: watching TV, playing video games, reading on laptops or tablets, or using smartphones: all of which are equipped with light-emitting screens.

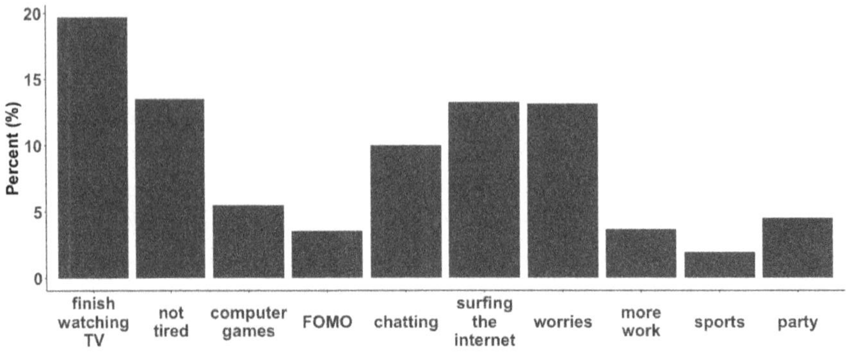

FIGURE 7.3 Reasons for not going to bed (N=1932)
SOURCE: FRAUNHOFER ISI/FOM CIRCADIA STUDY

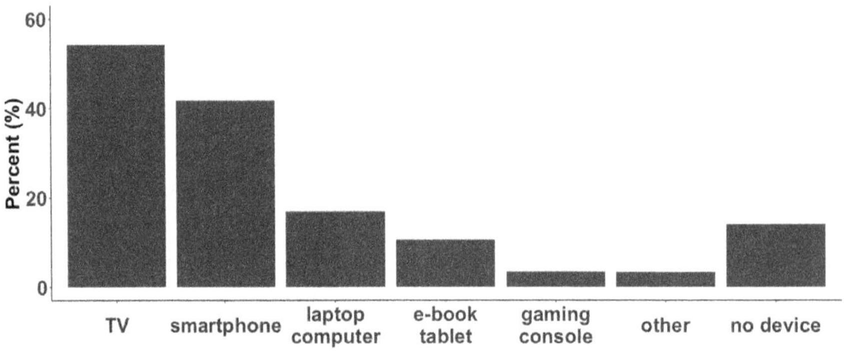

FIGURE 7.4 Device usage 1 hour before bedtime on workdays (N=1932)
SOURCE: FRAUNHOFER ISI/FOM CIRCADIA STUDY

It can thus be observed that the daily rhythms of people are influenced by devices more than they think. People not only use light-emitting devices for longer periods of time per day, but also at times when they need to calm down and when their bodies prepare for sleep. This can lead to less sleep-time and to serious consequences for health.

5 Outlook in Measuring Circadian Rhythms and Daily Life Activities

Our own research as well as that of others has revealed the critical importance of adapting life-style to circadian rhythms. A first and necessary step in this direction is that individuals identify their particular chronotype. One simple, tested instrument to assess this is the Munich ChronoType Questionnaire (MCTQ; Roenneberg et al. 2003; Kantermann, Sung, and Burgess 2015, Kantermann and

Eastman 2018), which has also been translated into a shorter version (Ghotbi et al. 2020). Chronotype is quantified by the MCTQ based on the mid-point of sleep on work-free days, corrected for sleep loss on workdays. The advantage of the MCTQ is that the result is a timepoint, which can easily be used to arrange work or school schedules. In addition, the MCTQ collects information about sleep duration, sleep inertia, sleep latency, use of alarm clocks and more for both workdays and work-free days. Another often-used instrument assessing a related topic, that of diurnal preference, is the Morningness/Eveningness questionnaire (MEQ, Horne and Ostberg 1976). The MEQ, however, gives only a final score, which is less useful for adjusting timetables.

Another, more objective but also more laborious method for judging chronotypes is through assessment of Dim Light Melatonin Onset from saliva or blood samples (DLMO; Kantermann and Eastman 2018; Kantermann, Sung, and Burgess 2015). The rise of melatonin in the evening prior to sleep varies in time and reflects an individual's chronotype. While a DLMO test is not available for home use, as it requires scientific staff support. Newly developed test called 'BodyClock' is offered on the German market for use at home. This test employs genetic testing of RNA activity in hair follicles and provides customers with information about their chronotype, in addition to science-based lifestyle recommendations.

Our research has led us to regard it as essential that people know about their individual chronotype and that companies know about the chronotype distribution of their staff. With the tools available, it is possible for employers, insurance agencies, and medical doctors to find out which chronotype group is most challenged in what situation. We all have a deep interest in doctors, nurses, fire workers, police officers, and pilots being alert and healthy while at work. The negative consequences of disregarding these challenges are significant. Together we can help mitigate the dilemmas that have been created by the current tension between societal life (school start, shift work, daylight saving time, etc.) and one's personal circadian rhythm. It's time to take chronobiology seriously, and tackle this next frontier of health literacy for the benefit of both the individual and society.

References

Arendt, Josephine. 2010. "Shift Work: Coping with the Biological Clock." *Occupational Medicine* 60 (1): 10–20. https://doi.org/10.1093/occmed/kqp162.

Basner, Mathias, and David F. Dinges. 2009. "Dubious Bargain: Trading Sleep for Leno and Letterman." *Sleep* 32 (6): 747–52. https://doi.org/10.1093/sleep/32.6.747.

Basner, Mathias, Andrea M. Spaeth, and David F. Dinges. 2014. "Sociodemographic Characteristics and Waking Activities and Their Role in the Timing and Duration of Sleep." *Sleep* 37 (12): 1889–1906. https://doi.org/10.5665/sleep.4238.

Cajochen, Christian, Sylvia Frey, Doreen Anders, Jakub Späti, Matthias Bues, Achim Pross, Ralph Mager, Anna Wirz-Justice, and Oliver Stefani. 2011. "Evening Exposure to a Light-Emitting Diodes (LED)-Backlit Computer Screen Affects Circadian Physiology and Cognitive Performance." *Journal of Applied Physiology* 110 (5): 1432–38. https://doi.org/10.1152/japplphysiol.00165.2011.

Costa, Giovanni. 2003. "Shift Work and Occupational Medicine: An Overview." *Occupational Medicine* 53 (2): 83–88. https://doi.org/10.1093/occmed/kqg045.

Cuhls, Kerstin, Thomas Kantermann, Friederike Mork-Antoni, Nils Heyen, and Ralph Gutknecht. 2024. *Chronobiologie für eine gesunde Gesellschaft der Zukunft. Projekt CIRCADIA—Circadiane Rhythmen und Technologie*. Karlsruhe.

Czeisler, C. A., J. F. Duffy, T. L. Shanahan, E. N. Brown, J. F. Mitchell, D. W. Rimmer, J. M. Ronda, et al. 1999. "Stability, Precision, and Near-24-Hour Period of the Human Circadian Pacemaker." *Science* 284 (5423): 2177–81. https://doi.org/10.1126/science.284.5423.2177.

Deutsche Gesetzliche Unfallversicherung. 2018. "DGUV Information 215–220 "Nicht-visuelle Wirkungen Von Licht Auf Menschen"." Unpublished manuscript, last modified November 17, 2021.

Ghotbi, Neda, Luísa K. Pilz, Eva C. Winnebeck, Céline Vetter, Giulia Zerbini, David Lenssen, Giovanni Frighetto, et al. 2020. "The MMCTQ: An Ultra-Short Version of the Munich ChronoType Questionnaire." *Journal of Biological Rhythms* 35 (1): 98–110. https://doi.org/10.1177/0748730419886986.

Hafner, Marco, Martin Stepanek, Jirka Taylor, Wendy Troxel, and Christian Stolk. 2017. *Why Sleep Matters: The Economic Costs of Insufficient Sleep*. RAND Corporation. Accessed January 28, 2022, https://pubmed.ncbi.nlm.nih.gov/28983434/.

Hale, Lauren, Gregory W. Kirschen, Monique K. LeBourgeois, Michael Gradisar, Michelle M. Garrison, Hawley Montgomery-Downs, Howard Kirschen, Susan M. McHale, Anne-Marie Chang, and Orfeu M. Buxton. 2018. "Youth Screen Media Habits and Sleep: Sleep-Friendly Screen Behavior Recommendations for Clinicians, Educators, and Parents." *Child and Adolescent Psychiatric Clinics of North America* 27 (2): 229–45. https://doi.org/10.1016/j.chc.2017.11.014.

Hansen, Johnni, and Christina F. Lassen. 2012. "Nested Case-Control Study of Night Shift Work and Breast Cancer Risk Among Women in the Danish Military." *Occupational and Environmental Medicine* 69 (8): 551–56. https://doi.org/10.1136/oemed-2011-100240.

Hastings, M. H., E. S. Maywood, and A. B. Reddy. 2008. "Two Decades of Circadian Time." *Journal of Neuroendocrinology* 20 (6): 812–19. https://doi.org/10.1111/j.1365-2826.2008.01715.x.

Hattar, S., H. W. Liao, M. Takao, D. M. Berson, and K. W. Yau. 2002. "Melanopsin-Containing Retinal Ganglion Cells: Architecture, Projections, and Intrinsic Photosensitivity." *Science* 295 (5557): 1065–70. https://doi.org/10.1126/science.1069609.

Horne, J. A., and O. Ostberg. 1976. "A Self-Assessment Questionnaire to Determine Morningness-Eveningness in Human Circadian Rhythms." *International Journal of Chronobiology* 4 (2): 97–110.

Kantermann, Thomas, and Charmane I. Eastman. 2018. "Circadian Phase, Circadian Period and Chronotype Are Reproducible over Months." *Chronobiology International* 35 (2): 280–88. https://doi.org/10.1080/07420528.2017.1400979.

Kantermann, Thomas, Myriam Juda, Martha Merrow, and Till Roenneberg. 2007. "The Human Circadian Clock's Seasonal Adjustment Is Disrupted by Daylight Saving Time." *Current Biology: CB* 17 (22): 1996–2000. https://doi.org/10.1016/j.cub.2007.10.025.

Kantermann, Thomas, Haein Sung, and Helen J. Burgess. 2015. "Comparing the Morningness-Eveningness Questionnaire and Munich ChronoType Questionnaire to the Dim Light Melatonin Onset." *Journal of Biological Rhythms* 30 (5): 449–53. https://doi.org/10.1177/0748730415597520.

Kantermann, Thomas, Sophie M. T. Wehrens, Melissa A. Ulhôa, Claudia Moreno, and Debra J. Skene. 2012. "Noisy and Individual, but Doable: Shift-Work Research in Humans." *Progress in Brain Research* 199: 399–411. https://doi.org/10.1016/B978-0-444-59427-3.00022-8.

Minors, D. S., J. M. Waterhouse, and A. Wirz-Justice. 1991. "A Human Phase-Response Curve to Light." *Neuroscience Letters* 133 (1): 36–40. https://doi.org/10.1016/0304-3940(91)90051-t.

Münch, Mirjam, Anna Wirz-Justice, Steven A. Brown, Thomas Kantermann, Klaus Martiny, Oliver Stefani, Céline Vetter, Kenneth P. Wright, Katharina Wulff, and Debra J. Skene. 2020. "The Role of Daylight for Humans: Gaps in Current Knowledge." *Clocks & Sleep* 2 (1): 61–85. https://doi.org/10.3390/clockssleep2010008.

Ritonja, Jennifer, Kristan J. Aronson, Raymond W. Matthews, Diane B. Boivin, and Thomas Kantermann. 2019. "Working Time Society Consensus Statements: Individual Differences in Shift Work Tolerance and Recommendations for Research and Practice." *Industrial Health* 57 (2): 201–12. https://doi.org/10.2486/indhealth.SW-5.

Roenneberg, Till, Tim Kuehnle, Myriam Juda, Thomas Kantermann, Karla Allebrandt, Marijke Gordijn, and Martha Merrow. 2007. "Epidemiology of the Human Circadian Clock." *Sleep Medicine Reviews* 11 (6): 429–38. https://doi.org/10.1016/j.smrv.2007.07.005.

Roenneberg, Till, Eva C. Winnebeck, and Elizabeth B. Klerman. 2019. "Daylight Saving Time and Artificial Time Zones—a Battle Between Biological and Social Times." *Frontiers in Physiology* 10: 944. https://doi.org/10.3389/fphys.2019.00944.

Roenneberg, Till, Anna Wirz-Justice, and Martha Merrow. 2003. "Life Between Clocks: Daily Temporal Patterns of Human Chronotypes." *Journal of Biological Rhythms* 18 (1): 80–90. https://doi.org/10.1177/0748730402239679.

van der Vinne, Vincent, Giulia Zerbini, Anne Siersema, Amy Pieper, Martha Merrow, Roelof A. Hut, Till Roenneberg, and Thomas Kantermann. 2015. "Timing of Examinations Affects School Performance Differently in Early and Late Chronotypes." *Journal of Biological Rhythms* 30 (1): 53–60. https://doi.org/10.1177/0748730414564786.

Zerbini, Giulia, Vincent van der Vinne, Lana K. M. Otto, Thomas Kantermann, Wim P. Krijnen, Till Roenneberg, and Martha Merrow. 2017. "Lower School Performance in Late Chronotypes: Underlying Factors and Mechanisms." *Scientific Reports* 7 (1): 4385. https://doi.org/10.1038/s41598-017-04076-y.

CHAPTER 8

The Circadian Clock and Circadian Rhythm Measurement

Ritsuko Matsumura

Abstract

The circadian clock, the internal system that allows various physiological functions to work properly at their appropriate times, has an almost, but not quite 24-hour cycle, as implied by the name "circadian," from the Latin "circa," "about," and "diem," "day." The actual substance of the circadian clock lies in the rhythm of increase and decrease within this period in the expression of clock genes, the genes that make up the molecular mechanism of the circadian clock. Accordingly, chronobiologists measure circadian rhythms in mammals by quantifying the expression levels of clock genes, allowing 'clock genes' or 'them' to investigate the functions of the circadian clock. The cycle is not only variable, only approximating a 24-hour period, but may also contain multiple redundant systems. Our research has shown that the transcriptional circadian rhythm of the clock gene *Cry*, which was considered important for the maintenance of circadian clock function, is in fact not essential in mice. This implies that there may be molecular mechanisms that maintain the circadian clock other than the transcriptional oscillations of clock genes, a redundancy like that built-in into many biological systems, where there are several routes to any given function or action rather than a single causal pathway.

Keywords

circadian rhythm – circadian clock – clock genes – transcriptional feedback loop – Cryptochrome (Cry) gene

The various physiological functions of our body follow a daily cycle controlled by an internal system known as the circadian clock. Body temperatures drop before we go to sleep and rise when we wake up (Aschoff 1983), while melatonin secretion increases during sleep and decreases during the day (Cajochen, Kräuchi and Wirz-Justice 2003), both due to the circadian clock. The circadian

 | DOI:10.1163/9789004723634_010

clock also influences our behavior: we rest at night and are active during the day.

The circadian clock influences biochemical, physiological, and behavioral processes, enabling various physiological functions to work properly at the appropriate times. This makes the functions more efficient and better able to aid the survival of the organism by anticipating and adapting to environmental changes. For this reason as well a disruption of the clock's function, as of the alignment between internal and external rhythms, or a disruption in rhythm period and amplitude, can affect our health. A disruption of phase alignment in particular, like that caused by staying up late or following irregular eating patterns, has been linked to diseases such as diabetes (Kurose, Yabe and Inagaki 2011), obesity (Albrecht 2017), depression (McClung 2007), and other sleep disorders (Sack et al. 2007). Disturbances in circadian rhythm are also known to increase the risk of cancer, mood disorders and other neurologic, psychiatric, metabolic, cardiovascular, and allergic/immunologic diseases (Roenneberg and Merrow 2016). Understanding how circadian rhythms function on a molecular level and what occurs when circadian rhythms are disrupted can help identify factors to target for treating multiple diseases and disorders.

Scientists have identified multiple genes that are responsible for producing circadian rhythms. In 2017, Jeffrey C. Hall and colleagues were awarded the Nobel Prize in Physiology or Medicine for identifying the "clock genes" in Drosophila that constitute the circadian clock and for discovering the molecular mechanisms that control circadian rhythms (Konopka and Benzer 1971). Clock genes have since also been discovered in mammals (Vitaterna et al. 1994, Tei et al. 1997), plants, and even bacteria (Pruneda-Paz et al. 2001, Ishiura et al. 1998). Before presenting data from our own research, therefore, we will explore the circadian clock's not-quite 24-hour cycle on a molecular level.

1 Introduction to the Circadian Clock

In contrast to the circadian clock, the word "clock" implies a device that has regular movements in precisely equal intervals. Mechanical clocks can keep time through regular oscillations of a pendulum or spring, displaying the time of day in the form of hours and minutes. The important point in this is that they oscillate regularly. With this understanding, it can be said that the circadian clock is an internal mechanism that produces an oscillation of an approximately one day period. And while the time resets through exposure to light every day, so that it roughly follows the alternation of day and night, the

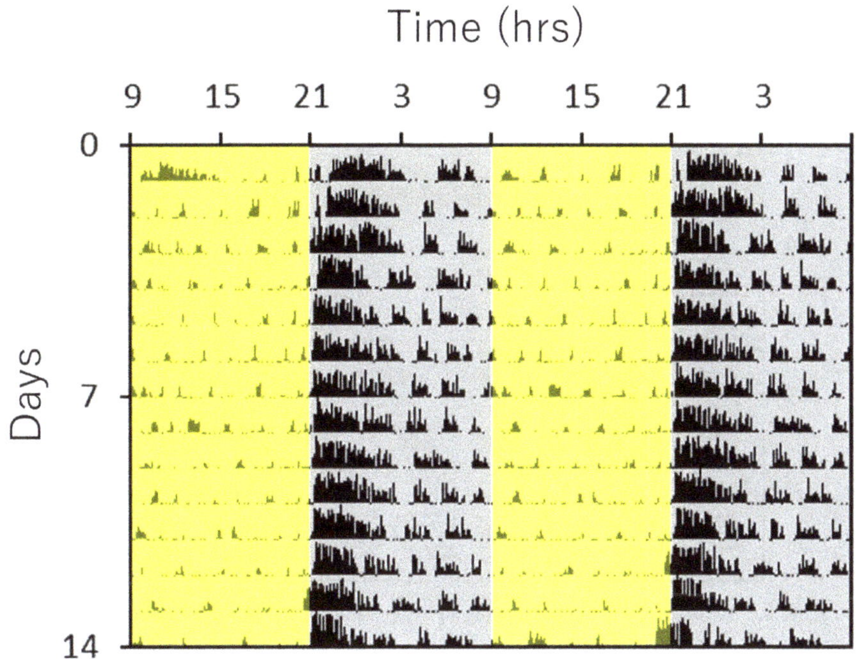

FIGURE 8.1 Actogram of a mouse showing 14 days of activity

circadian clock is in fact determined by an internal oscillation rhythm produced by molecules, genes and proteins. This is significant since changes at the molecular level are reflected in physiological functions or behavior. Similarly, measurement of the influence on behavior on a larger scale is also important to circadian clock research at the molecular level.

Behavioral rhythms can be charted in an "actogram" such as Figure 8.1, in which the locomotor activity level of mice is indicated by the height of the bars. The horizontal axis indicates the time of day over two days, while the yellow columns show the light period and the gray columns the dark period during these days, namely 12 light hours followed by 12 dark hours. Consistent with the fact that mice are nocturnal, this graph shows that they are more active in the dark period. In contrast, we humans are diurnal, which means that we are active during the day and rest at night. But while Figure 8.1 might appear to show that mice are simply responding to the change from day to night, these changes are not simply a response to light. This is shown by the fact that the circadian rhythm persists even without the change from light to dark, as seen in Figure 8.2, where mice locomotion is tracked in conditions of continual darkness.

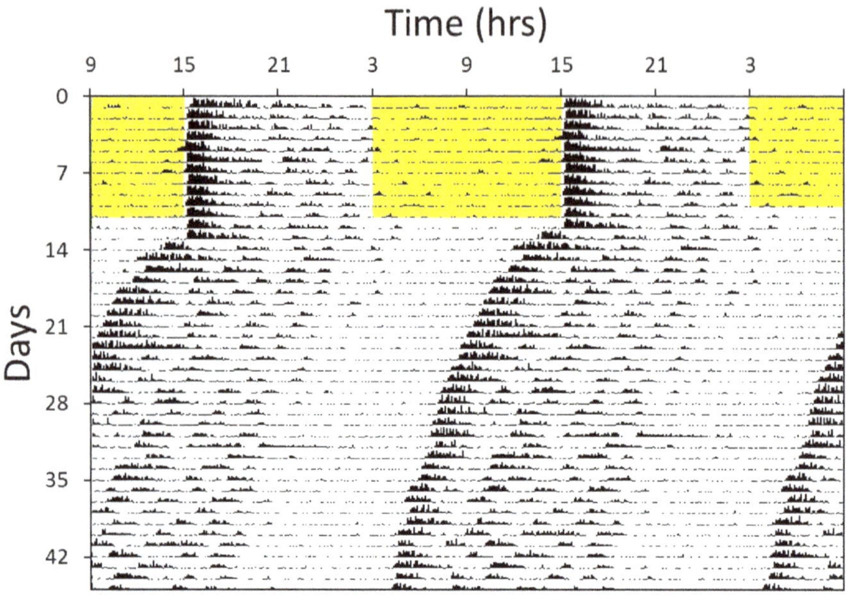

FIGURE 8.2 Actogram of a mouse showing 34 days of activity in constant darkness following 11 days in a light–dark condition

Figure 8.2 shows an actogram of a mouse after the original alternation of light and dark was changed to a condition of constant darkness. As shown, the circadian rhythm continues even under the constant-dark condition. This chart also makes it clear that the circadian clock is regulated not simply by the alternation of light and dark, but rather is driven endogenously, in other words, at a molecular level. Such a rhythm under constant conditions is called a "free-running" rhythm, and reflects the endogenous, internal or "natural" period of the circadian clock in the absence of external light cues. In this example, the free-running period is close to, but not exactly 24 hours. This mouse's free-running period is a little shorter than 24 hours, so that the mouse's active period has gradually shifted to the left, indicating that the mouse begins to be active earlier each subsequent day.

2 Mechanisms of the Circadian Clock

In order to explain the significance of our research, showing that the circadian rhythm of transcription of the *Cry1* gene, an element that had been thought essential for maintaining circadian clock function, is not in fact needed in mice, it will be necessary to go into some further scientific background. The main clock genes which make the mammalian circadian clock tick are *Bmal1*, *Clock*, *Period1* and *2* (*Per1, 2*), and *Cryptochrome1* and *2* (*Cry1, 2*). The paper published about 50 years ago on a study in a small fly, Drosophila, showed that the circadian rhythm of the fly's hatching is controlled by a single gene (Konopka and Benzer 1971) named Period, which, together with its molecular mechanism, was later discovered in mammals as well (Tei et al. 1997; Kume et al. 1999). Mutations in this gene caused the circadian rhythm to disappear or the period length to change.

2.1 *Molecular Mechanism of the Circadian Clock*

The circadian clock mechanism consists of a feedback loop of clock gene expression (Kume et al. 1999), a process in which the outputs of a system are circled back and used as inputs. During gene expression, that is, the production of proteins from information carried in genes, proteins are produced through two steps: transcription and translation (Figure 8.3). In the first step, transcription, the genetic information is copied from DNA, the gene, to a molecule called mRNA.[1] Based on the information contained in the mRNA, amino acids are linked together in sequence to form proteins. This step is called translation. An autoregulatory feedback loop based on this gene expression process is the basic mechanism of the circadian clock. The details of the mechanism are explained as follows. Figure 8.4 shows a single cell. Inside the cell is a structure called the nucleus, in which genes are stored. The portion surrounding the nucleus is called the cytoplasm. Figure 8.5 explores the molecular circadian clock mechanism of the cell, showing *Per* and *Cry* genes inside the nucleus as well as BMAL1 and CLOCK proteins which promote their transcription. The presence of BMAL1-CLOCK results in the rapid production of PER and CRY which interact in the cytoplasm and migrate into the nuclei, suppressing the function of BMAL1-CLOCK.[2] The production of PER and CRY accordingly decreases, as does the suppression, which in turn makes the expression of PER and CRY increase again. This loop is called a transcription-translation negative feedback loop. It takes about one day for the loop to complete one cycle.

The circadian clock is thus essentially the rhythm of the increase and decrease in the expression of clock genes in the cell.

The clock genes are then involved in the regulation of genes that affect physiological functions, such as the secretion of hormones, and so influence our behavior on the macro level. At the same time, the circadian clock mechanism

---

1 Messenger RNA. or mRNA, is a molecule that contains and coveys the instructions which direct cells to make a specific protein.

2 Italics refers to the gene (DNA sequence), while the capitalized form refers to the protein.

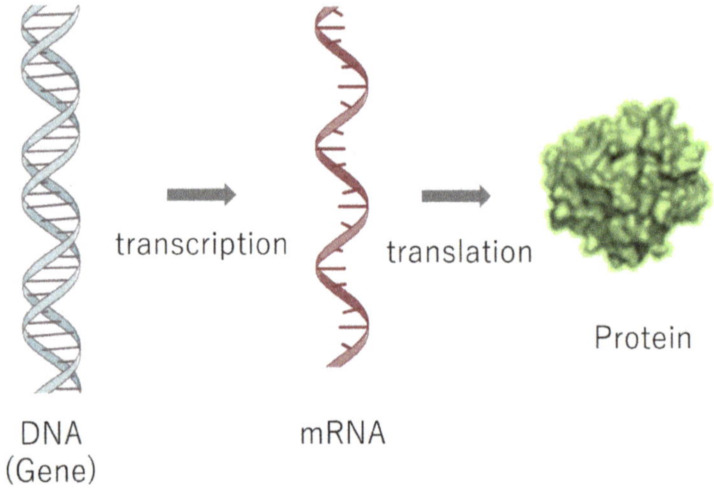

FIGURE 8.3 Gene expression undergoing two steps: transcription and translation

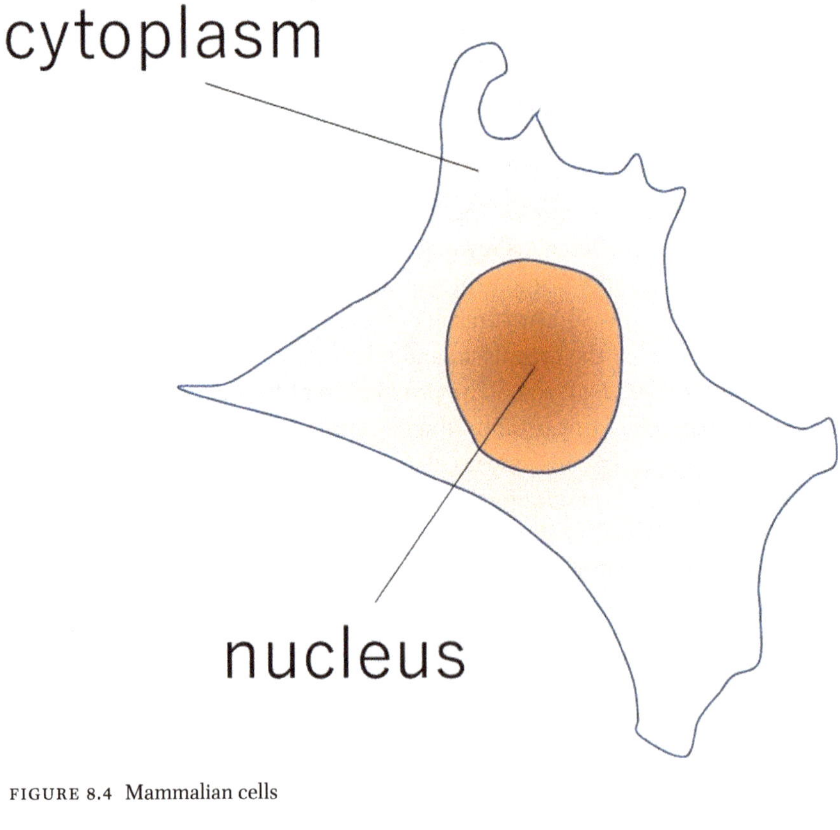

FIGURE 8.4 Mammalian cells

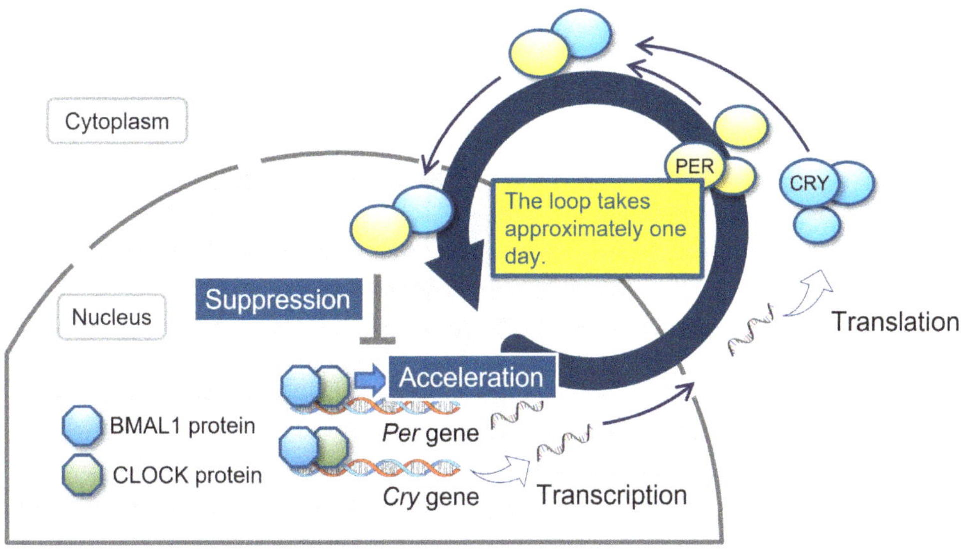

FIGURE 8.5 Schematic diagram of the molecular model of the mammalian circadian clock, showing the transcription-translation feedback loop

occurs within a single cell. So, it is often said that "every single cell has one circadian clock."

2.2 *The Circadian Clock System at the Whole-Body Level*

While the molecular mechanism of the clock genes occurs within a single cell, the circadian clock system at the whole-body level requires the existence of a master clock, which exists in the brain and coordinates the circadian clocks at the tissue and cellular levels. In humans and other mammals, the central clock is in the SCN (suprachiasmatic nucleus) within the hypothalamus in the brain (Stephan and Zucker 1972; Moore and Eichler 1972). In mammals the SCN orchestrates the circadian clocks throughout the body, called peripheral clocks, like an orchestra conductor, or a radio tower that sets the time on radio-controlled clocks. It is this master clock that resets circadian rhythms based on light stimuli (Czeisler 1995), the effects of light being significantly different depending on time of day. The SCN thus adjusts the circadian clock so that it is not out of phase with the environment. However, since we cannot respond immediately to a sudden change in light stimuli (Leloup and Goldbeter 2013), it can take the body several days to adjust in the case of extreme changes, which is why, for example, we experience jet lag, an occasion on which we directly experience the presence of the circadian clock. The circadian clock system is summarized in Figure 8.6.

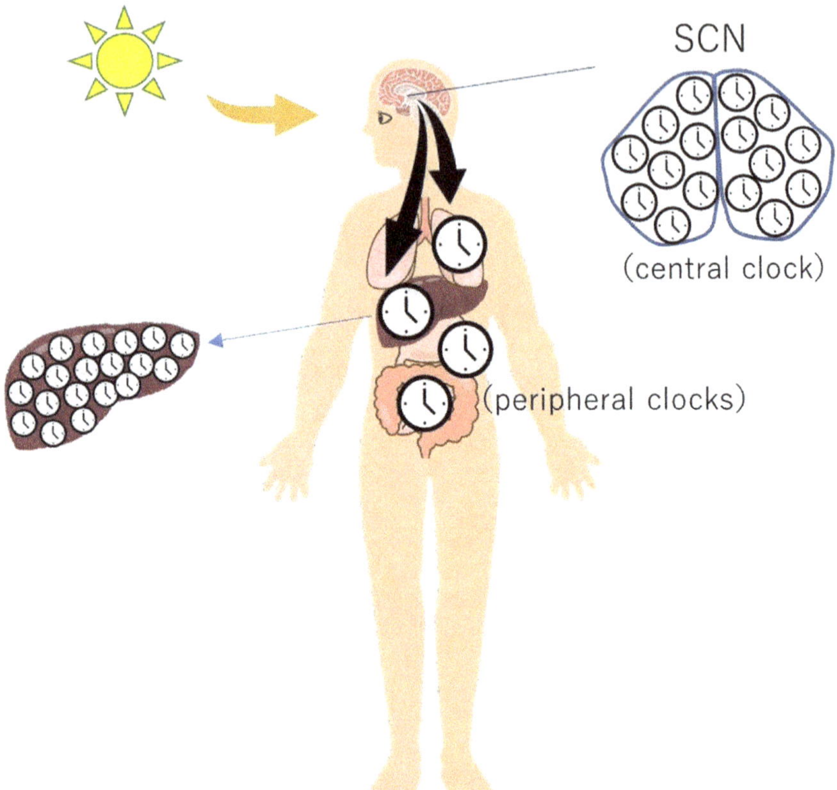

FIGURE 8.6 The circadian clock system of mammals

3 The Measurement of Circadian Rhythm

Measuring the rhythm of clock gene expression is also useful in assessing the state of the circadian clock, as in our own research. This is because, as described above, the fundamental characteristic of the circadian clock is the rhythm of expression of the clock genes themselves.

As the circadian clock is fundamentally a feedback loop composed of gene expression processes, the circadian rhythm can be measured by a time course analysis of mRNA or protein, which are produced during gene expression. There are two main methods, both of which we have used in our research. One is the direct quantitative analysis of mRNA or protein levels at regular intervals for a period of one or more days. The second is an indirect quantitative analysis using reporter genes, genes that can be both attached to other elements and easily measured. Direct quantitative analysis methods can measure multiple genes at once and require no genetic manipulation, but are very laborious. The

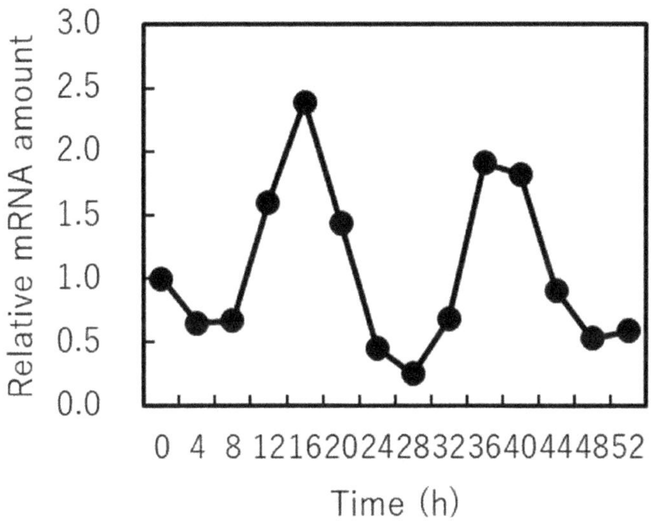

FIGURE 8.7 Data from a direct quantitative analysis of mRNA of clock genes

indirect quantitative analysis method, using reporter genes, allows for high temporal resolution and is less laborious, but requires genetic manipulation.

3.1 *Direct Quantitative Analysis Methods*

Experimental data from a direct quantitative analysis are shown in Figures 8.7 and 8.8. The mRNA is quantified by a method called real-time quantitative polymerase chain reaction, or RT-qPCR.[3] Figure 8.7 shows data from a mouse tissue sample in which mRNA for the clock gene was sampled for two days at 4-hour intervals, showing that the amount of mRNA increases and decreases over an approximately 24-hour cycle. Figure 8.8 shows detection of the Per protein, quantified through a method called Western blotting.[4] Western blotting allows for the visualization of a protein of interest that can then be quantified. We can see that the PER also increases and decreases in a cycle of about 24 hours.

3 Polymerase Chain Reaction (PCR) is a technique used in molecular biology to make copies of a specific segment of DNA. It is like a molecular photocopying machine for DNA, and allows scientists to analyze and manipulate specific DNA sequences with high precision. Real-time PCR is one of the most widely used methods for analyzing gene expression. qPCR requires converting the RNA into DNA (called cDNA), and then measuring the resulting cDNA concentration.

4 Western blotting is a research tool used to identify and quantify specific proteins in biological samples. The proteins are first separated and transferred to a filter. Antibodies that stick to the target protein are then added and a marker is used to see where these proteins are on the filter.

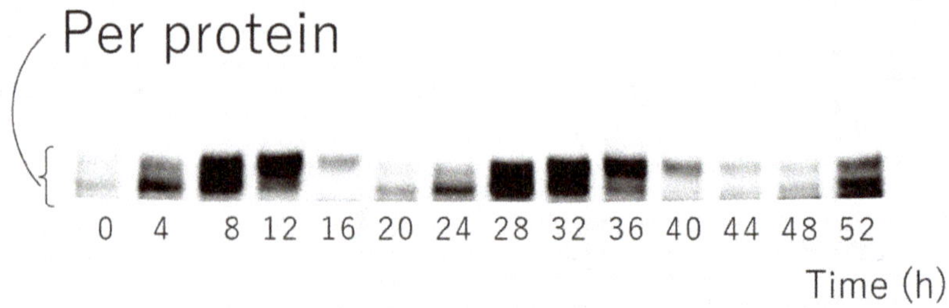

FIGURE 8.8 Data from a direct quantitative analysis of proteins produced from clock genes

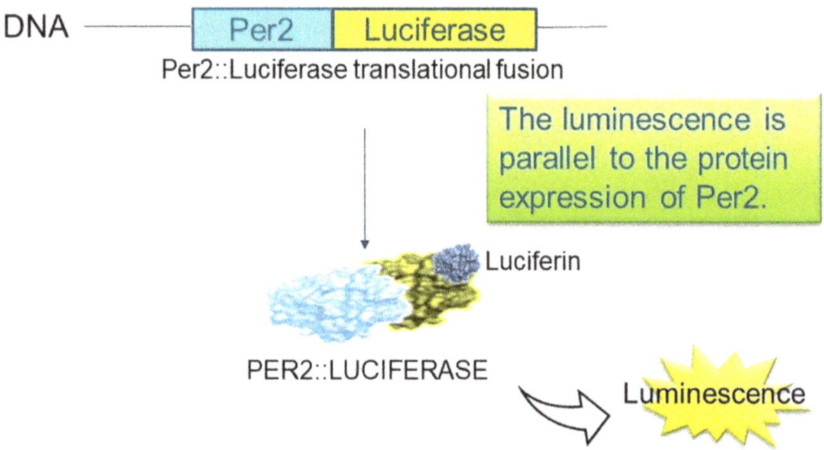

FIGURE 8.9 Luminescence occurs in parallel with production of the Per2 protein because the Per2-fused luciferase catalyzes a chemical reaction in which the substrate, luciferin, reacts and emits light

3.2 *Indirect Quantitative Analysis Method*

Indirect quantitative analysis uses a reporter gene, here luciferase which produces bioluminescence, to quantify the rhythm of clock gene expression. The firefly luciferase gene, an enzyme that catalyzes the reaction of bioluminescence, is often employed. This luciferase gene is genetically engineered to fuse with the target gene in a process called translational fusion, illustrated in Figure 8.9. DNA is artificially modified so that *Per2* and *luciferase* are linked, creating a fusion protein of *Per2* and *luciferase* (PER2-LUC).

Here luciferase catalyzes the chemical reaction in which luciferin emits light, a luminescence that corresponds to the production of the PER2. By measuring this luminescence, we can infer the levels of PER2 at any given time, and since data can be acquired at intervals of a few minutes, the technique provides data with a high temporal resolution.

4 Research Example[5]

4.1 *Description of the Research and Background*

"Is the oscillation of Cry gene transcription essential for the circadian clock machinery in mice?" This is the essential question of our research project. In this study, the significance of the oscillation of *Cry* gene transcription, assigned an important role in the current molecular model of the circadian clock, was examined using the molecular biological techniques described above. To conduct this research, we first identified *Cry* gene-specific cis-elements (regulatory DNA that controls *Cry* gene circadian transcription) and then produced mice lacking a cis-element in order to assess the circadian clock function of the mice possessing an altered *Cry* gene.

We identified the cis-elements that are responsible for the transcriptional oscillation of the *Cry* gene using genome-edited cultured mouse cells, that is, a mouse cell line, NIH 3T3, grown in a petri dish. Cis-elements are particular regions on DNA where specific other proteins involved in the initiation and regulation of transcription bind. These are mostly located in the promoter region, a DNA region located upstream of a gene that controls the quantity and

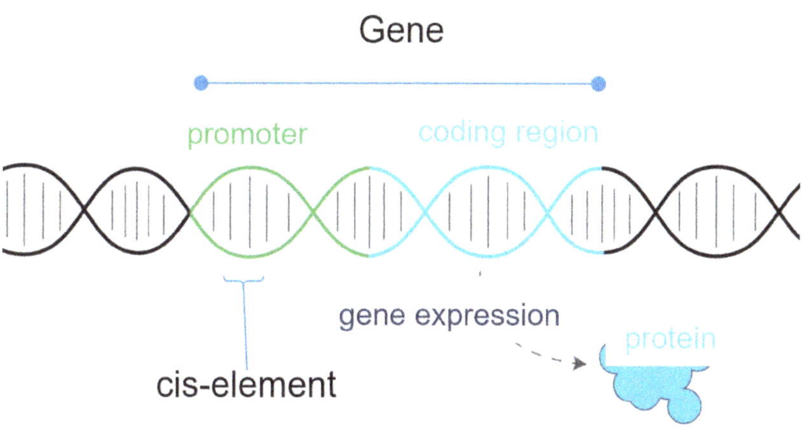

FIGURE 8.10 Schematic diagram showing the positional relationship of genes, promoters, and cis-elements

5 This study was published in Cell Reports 39 (3) 2022, so if you would like to learn more details, please refer to it as well.

timing of the gene's expression. Figure 8.10 shows their positional relationship. Our previous research had found a cis-element essential for the transcriptional circadian oscillation of *Cry*, which we named E-box, as its DNA sequence was nearly identical to E-box, a popular cis-element. By disrupting this cis-element we were able, essentially, to turn off the circadian rhythm of *Cry* transcription.

4.2 *Summary of Research Results*

We disrupted E-box in the genome of cultured mouse cells by genome editing. The transcription rhythm of clock genes was then examined by directly quantifying mRNA. We found that the transcription of the clock gene *Bmal1* still had a circadian rhythm, but as expected, no overt rhythm was found in *Cry* (Figure 8.11). This allowed us to show that the E-box is endogenously responsible for the transcriptional oscillation of *Cry*.

Next, we assessed circadian clock function at the whole-body, tissue, and cell levels using gene-altered mice lacking the *Cry* transcriptional rhythm. We generated mice that had disrupted *Cry1* and *Cry2* E-boxes using genome editing. The *Per2-luciferase* gene was also introduced so that the circadian rhythm could be measured in real-time monitoring as bioluminescence (Figure 8.12). Circadian locomotor activity rhythms were first examined.

The results of our examination revealed that while the mice with altered *Cry* genes exhibited free-running circadian rhythms, showing that *Cry* was not essential for generating a circadian rhythm, those rhythms were nonetheless affected by the change and differed from normal, or "wild-type," mice. The gene-altered, or "mutant," mice showed overt circadian rhythms in free-running activity but also had a longer circadian period than normal, or wild-type mice (Figure 8.13). In wild-type mice, the free-running period was a little shorter than 24 hours, so that the active phase gradually shifted to the left. In mutant mice, however, the active phase shifted to the right showing that the

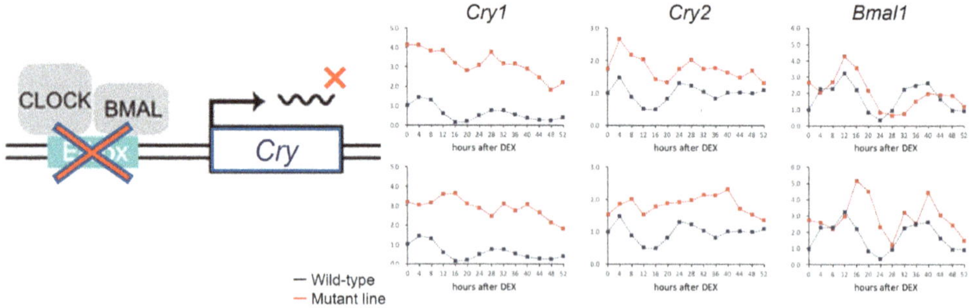

FIGURE 8.11 The transcription rhythm of clock genes in the cultured mouse cells, in which the cis-element of clock genes were disrupted, examined by directly quantifying mRNA

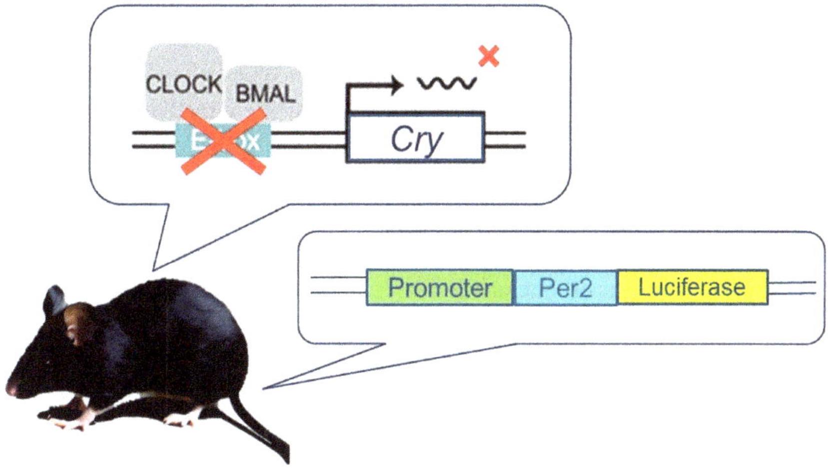

FIGURE 8.12 The mice we generated had disrupted *Cry1* and *Cry2* E-boxes, which caused the loss of the overt circadian transcriptional rhythm of *Cry* genes

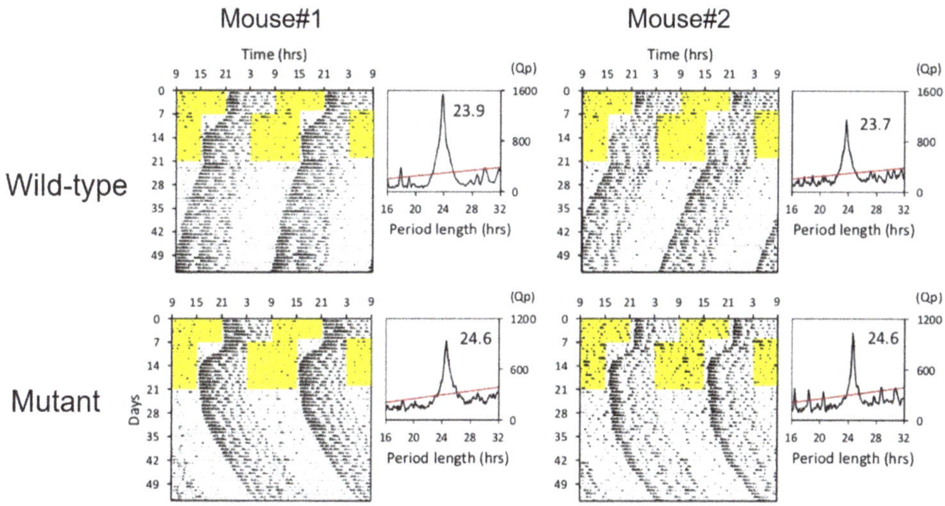

FIGURE 8.13 Actograms of wild-type and mutant mice

free-running period here was longer than 24 hours, and so also longer than that of the wild-type mice.

We then examined the tissue-level circadian clock function of the mutant mice, with the same result. The circadian rhythm of tissue can be measured using slice culture, in which slices of the tissues were cultured and luminescence measured. As the mutant mice have the *Per2-luciferase* fusion gene, we could measure the circadian rhythm of *Per2* expression as changes in luminescence. The mutations caused no severe defects in tissue-autonomous circadian rhythm generation in any of the three types of tissue examined, namely liver, lung, and aorta (Figure 8.14). Again, the circadian rhythm also had a longer period than in the wild-type mice, so that the peak of the waveform was gradually shifted to the right (Figure 8.14a). For the lungs, we directly quantified mRNA and measured the levels over time. The results are shown in the graphs, Figure 8.14b, and indicate that *Cry* transcription lost its circadian rhythm, whereas other clock genes (*Per2*, *Bmal1*, *Rev-erbα*) showed a circadian rhythm.

Finally, we examined circadian rhythms at the cellular level in the mutant mice, finding, on this level, somewhat different results. Cells termed mouse embryonic fibroblasts (MEFs), were prepared from the mutant mice. Because these cells, of course, also had *Per2-luciferase* genes, we could measure their circadian rhythm by adding luciferin into the medium. Figure 8.15a shows the quantity of directly detected clock gene mRNA. As expected, *Bmal1* showed an overt circadian rhythm, but not only *Cry1*, *2* but also *Per2*, surprisingly, did not. In the rhythm of the bioluminescence (Figure 8.15b), a difference in circadian amplitude, and phase delay, due to longer period length between wild-type and mutant mice, was visually detectable. Excluding period length, this differs from the results in the tissues where a fairly normal circadian rhythm of PER2-LUC

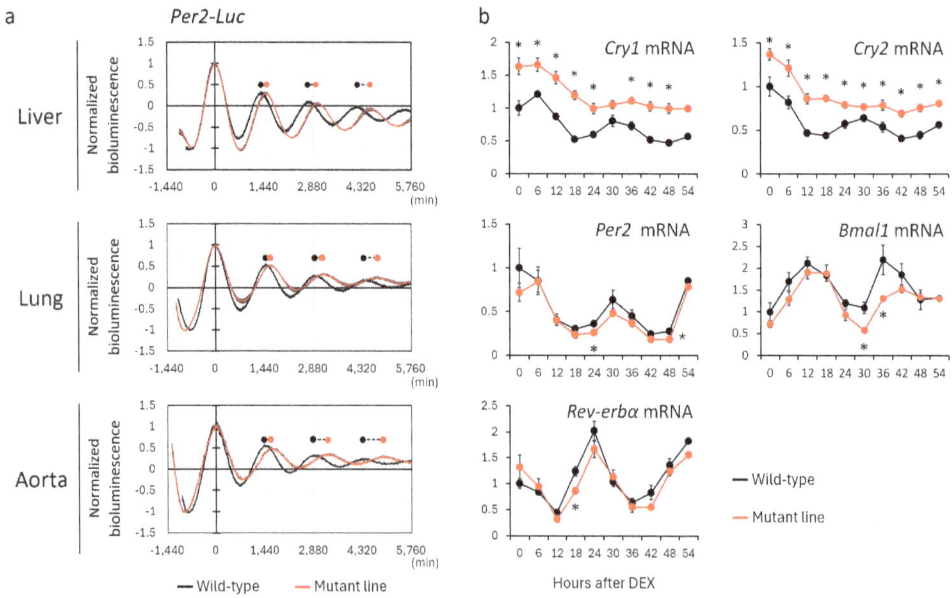

FIGURE 8.14 Disruption of the circadian transcriptional rhythm of *Cry* genes

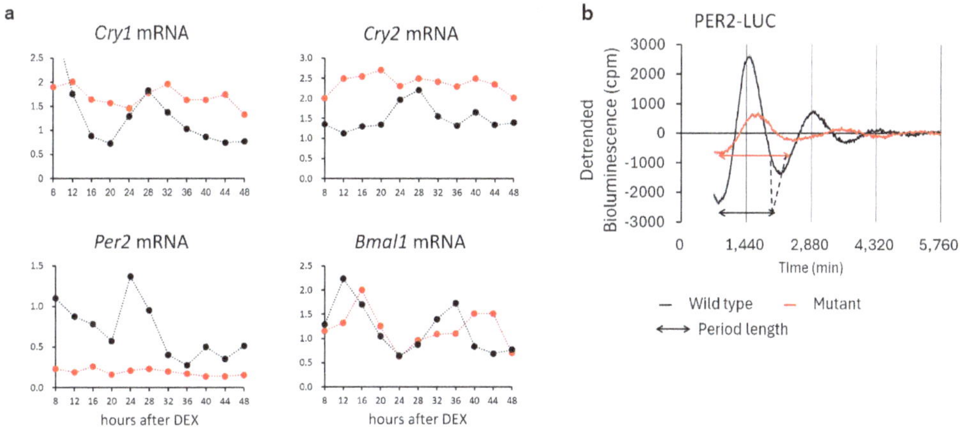

FIGURE 8.15 The circadian rhythms at the cellular level in mutant mice

was detected by real-time bioluminescence measurement (Figure 8.14a). These inconsistencies can be explained by the hypothesis that the different types of cells we used, that is, liver, lung, aorta cells, and MEFs, have different molecular aspects in the circadian oscillator and/or that tissue architecture increases the circadian robustness of the molecular machinery. Although we do not presently have any evidence to support these hypotheses, the latter hypothesis may be supported by a recent study demonstrating the existence of intercellular coupling of cell-autonomous circadian oscillators in the liver in vivo (Sinturel et al. 2021).

The conclusions of this study are therefore as follows:

1. Mice lacking *Cry* transcription oscillation nonetheless show robust circadian rhythms.
2. Tissue-autonomous circadian rhythms on the level purely of tissue are also robust without *Cry* oscillation.
3. While cell-autonomous *Cry* oscillation is dispensable for circadian rhythm, it contributes to tuning the circadian period length.

In short, this study showed that the transcriptional circadian rhythm of *Cry*, which had been considered important for the maintenance of circadian clock function, is in fact not essential. The implication follows that there may be molecular mechanisms maintaining the mammalian circadian clock other than transcriptional oscillations of clock genes, when we include the fact that overt *Per2* transcriptional rhythm at the cellular level, in which the *Cry* transcriptional circadian rhythm was disrupted, is also lost.

As described above, measuring the expression rhythms of clock genes gives us insight into the molecular components of the circadian clock, including an assessment of the role of the *Cry* circadian transcription rhythm. While not essential, suggesting that mechanisms other than the transcriptional oscillations of clock genes may maintain the mammalian circadian clock, *Cry* circadian transcription rhythm appears to contribute to tuning the circadian period length. Just as the discovery of clock genes and the development of molecular biological techniques during the last decades have made it possible to measure the circadian expression rhythms of clock genes, this in turn has led to progress in the ongoing study of the essential workings of the circadian clock.

References

Albrecht, Urs. 2017. "The Circadian Clock, Metabolism and Obesity." *Obesity Reviews* 18 (S1): 25–33. https://doi.org/10.1111/obr.12502.

Aschoff, Jürgen. 1983. "Circadian Control of Body Temperature." *Journal of Thermal Biology* 8 (1): 143–47. https://doi.org/10.1016/0306-4565(83)90094-3.

Cajochen, C., K. Kräuchi, and A. Wirz-Justice. 2003. "Role of Melatonin in the Regulation of Human Circadian Rhythms and Sleep." *Journal of Neuroendocrinology* 15 (4): 432–37. https://doi.org/10.1046/j.1365-2826.2003.00989.x.

Czeisler, Charles A. 1995. "The Effect of Light on the Human Circadian Pacemaker." In Ciba Found Symp 183: 254–90; discussion 290–302. https://doi.org/10.1002/9780470-514597.ch14.

Ishiura, Masahiro, Shinsuke Kutsuna, Setsuyuki Aoki, Hideo Iwasaki, Carol R. Andersson, Akio Tanabe, Susan S. Golden, Carl H. Johnson, and Takao Kondo. 1998. "Expression of a Gene Cluster kaiABC as a Circadian Feedback Process in Cyanobacteria." *Science* 281 (5382): 1519–23. https://www.science.org/doi/10.1126/science.281.5382.1519.

Konopka, Ronald J., and Seymour Benzer. 1971. "Clock Mutants of Drosophila Melanogaster." *Proceedings of the National Academy of Sciences* 68 (9): 2112–16. https://doi.org/10.1073/pnas.68.9.2112.

Kume, Kazuhiko, Mark J Zylka, Sathyanarayanan Sriram, Lauren P Shearman, David R Weaver, Xiaowei Jin, Elizabeth S Maywood, Michael H Hastings, and Steven M Reppert. 1999. "mCRY1 and mCRY2 Are Essential Components of the Negative Limb of the Circadian Clock Feedback Loop." *Cell* 98 (2): 193–205. https://doi.org/10.1016/S0092-8674(00)81014-4.

Kurose, Takeshi, Daisuke Yabe, and Nobuya Inagaki. 2011. "Circadian Rhythms and Diabetes." *Journal of Diabetes Investigation* 2 (3): 176–77. https://doi.org/10.1111/j.2040-1124.2011.00105.x.

Leloup, Jean-Christophe, and Albert Goldbeter. 2013. "Critical Phase Shifts Slow down Circadian Clock Recovery: Implications for Jet Lag." *Journal of Theoretical Biology* 333 (September): 47–57. https://doi.org/10.1016/j.jtbi.2013.04.039.

McClung, Colleen A. 2007. "Circadian Genes, Rhythms and the Biology of Mood Disorders." *Pharmacology & Therapeutics* 114 (2): 222–32. https://doi.org/10.1016/j.pharmthera.2007.02.003.

Moore, Robert Y., and Victor B. Eichler. 1972. "Loss of a Circadian Adrenal Corticosterone Rhythm Following Suprachiasmatic Lesions in the Rat." *Brain Research* 42 (1): 201–6. https://doi.org/10.1016/0006-8993(72)90054-6.

Pruneda-Paz, Jose L., Ghislain Breton, Alessia Para, and Steve A. Kay. 2001. "A Functional Genomics Approach Reveals CHE as a Component of the Arabidopsis Circadian Clock." *Science* 293 (5531): 880–83. https://doi: 10.1126/science.1061320.

Roenneberg, Till, and Martha Merrow. 2016. "The Circadian Clock and Human Health." *Current Biology* 26 (10): R432–43. https://doi.org/10.1016/j.cub.2016.04.011.

Sack, Robert L, Dennis Auckley, R. Robert Auger, Mary A. Carskadon, Kenneth P. Wright, Michael V. Vitiello, and Irina V. Zhdanova. 2007. "Circadian Rhythm Sleep Disorders: Part I, Basic Principles, Shift Work and Jet Lag Disorders An American Academy of Sleep Medicine Review." *Sleep* 30 (11): 1460–83. https://pmc.ncbi.nlm.nih.gov/articles/PMC2082105/.

Stephan, Friedrich K., and Irving Zucker. 1972. "Circadian Rhythms in Drinking Behavior and Locomotor Activity of Rats Are Eliminated by Hypothalamic Lesions." *Proceedings of the National Academy of Sciences* 69 (6): 1583–86. https://doi.org/10.1073/pnas.69.6.1583.

Sinturel, Flore, Pascal Gos, Volodymyr Petrenko, Claudia Hagedorn, Florian Kreppel, Kai-Florian Storch, Darko Knutti, Andre Liani, Charles Weitz, Yann Emmenegger, Paul Franken, Luigi Bonacina, Charna Dibner, and Ueli Schibler. 2021. "Circadian Hepatocyte Clock Keep Synchrony in the Absence of a Master Pacemaker in the Suprachismatic Nucleus or Other Extrahepatic Clocks." *Genes Dev.* 1; 35(5–6): 329–334. http://10.1101/gad.346460.120.

Tei, Hajime, Hitoshi Okamura, Yasufumi Shigeyoshi, Chiaki Fukuhara, Ritsuko Ozawa, Matsumi Hirose, and Yoshiyuki Sakaki. 1997. "Circadian Oscillation of a Mammalian Homologue of the Drosophila Period Gene." *Nature* 389 (6650): 512–16. https://doi.org/10.1038/39086.

Vitaterna, Martha Hotz, David P. King, Anne-Marie Chang, Jon M. Kornhauser, Phillip L. Lowrey, J. David McDonald, William F. Dove, Lawrence H. Pinto, Fred W. Turek, and Joseph S. Takahashi. 1994. "Mutagenesis and Mapping of a Mouse Gene, Clock, Essential for Circadian Behavior." *Science* 264 (5159): 719–25. https://doi.org/10.1126/science.8171325.

CHAPTER 9

The Body Narrated: Shifting Time Zones in Medieval Legends and Tales

Daniela Tan

Abstract

Body time is generally understood as a linear process of physical changes, but this succession is rearranged in the world of narration. This is inherent in folk tales and legends, which are filled with miracles and magic, but what then happens when linear chronobiological time meets the shifting time zones of the narrative world? This paper is about women who enter a forbidden territory, thereby awakening powerful and threatening forces. In particular, the sudden recurrence of menstruation is a characteristic shared by these stories. But while this has traditionally been interpreted as a punishment and heavenly threat, I argue that it can equally be seen as rejuvenation. This empowering act of regaining vital forces thus defies the normal measure of time, just as it accompanies the transgression of discriminating practice. The stories discussed are folk tales and Buddhist fables known as *setsuwa*, legends which have been orally transmitted and recorded in many variations. Among the elements that can be seen throughout all versions, is a a common motif of women who do not accept their exclusion from sacred areas because of their sex. I will discuss stories with this motif, from the oldest version recorded in a 11th century legend collection on Immortals (*Honchō shinsen den*) up to the 17th century Buddhist tale *Karukaya*. As we will see, over time, the protagonists become depicted as less powerful and courageous commensurate with a shift towards patriarchal practices in religion, politics and society during medieval Japan. While physical processes and timelines frame the stories, they are also broken by unconformities in temporal measurement. The diegetic world of narrative confronts the reader with fractures between the processes of chronobiological time and the course of events in the story: processes are accelerated or reversed, and temporal segments such as stages of maturity are skipped, inverted, or intensified. What narrative intention is revealed in analyzing this technique? Does it point to a deformation of time or is it rather an account of the constant shifting of human existence? Or do these inversion in measures of time serve to subvert standard patriarchal interpretations? Not only time, but also shape and gender, are subject to constant change in the world of tales and legends. This paper focuses on the time of the body within the magical world of fable, to show how the social construction of woman,

 | DOI:10.1163/9789004723634_011

shaped by a patriarchal measure, may also be exposed through a narrative rearrangement of time.

Keywords

body time – menstruation – Buddhist tales – menopause – medieval Japan – mountain religions – blood taboo – female immortals

1 Introduction[1]

The physical body as a material entity is a transient appearance, nothing more than a temporary form that comes into life and then moves continually toward death. Each shape, whether of human or animal, is different, yet we share some common features, such as a beating heart, digestive and reproductive organs, glands and circulatory organs, and most of all, a lifespan. We are born, we age, and we die. What could be more of a linear flow than this? However, the tales presented in this paper tell a different story,[2] as they feature post-menopausal women who experience a sudden onset of their menstrual bleeding. In these stories, the linear chronobiology of the physical body is inverted and rearranged for narrative reasons. This may not come as a surprise in the miraculous world of tales and legends. Dramatically, these narrative twists raise the tension for listeners and readers as they also form a pattern revealing a deeper layer of meaning. But while research on this topic has focused on female impurity, my analysis will argue instead that this traditional interpretation has been conditioned by Buddhism from the early 10th century onward, and by its positioning of the social status and position of women.

This paper introduces a new perspective, by considering blood as a signifier of life force rather than of defilement, thus adding a new focus on female vital powers. By focusing on women and the physical processes of reproduction in

1 This paper has received funding from the European Research Council (ERC) under the European Union's Horizon 2020 research and innovation program (grant agreement No. 741166).

2 This view of the human body is reflected in the word *karada* 体, which suggests a continuity with the homonymous word *kara* 殻 (shell, husk), and also resonates with *kara* 空 (emptiness). The term *shintai* 身体 or *mi* 身 on the other hand is used in a more neutral sense to designate the social body of a person, and one's position in society, as exemplified in *mibun* 身分. Let's use the terms *nikutai* 肉体 and *shintai* 身体 for the sensual and societal aspects of the human body.

relation to a linear measure of time, the analysis of the stories will show that the temporal rearrangement exhibited in the diegetic world reveals another view of the female body and its reproductive capacities, now as a subject of worship.

The stories discussed in this paper share the motif of a woman entering a forbidden area and subsequently awakening powerful forces. As many versions exist, it can be assumed that the shape of the story has changed over time, absorbing the worldview and mood of each time of its telling. As will be shown, the women in the older versions act more boldly, refusing to surrender to a law that cuts them off from what they consider their natural right. By refocusing on this aspect of the stories, I argue for an alternative reading, that these stories present us with the thought that certain individuals may enter sacred sites and release vital powers—whether they are male or female.

I will draw from genres located in the intermediate zone between orally transmitted literature and written sources, such as legends (*setsuwa* 説話), folk tales, and Buddhist sermon tales (*sekkyō* 説経),[3] all of which bear traces of stories told over and over again over many centuries. As stories conveyed largely through an oral rather than written tradition, these tales may be regularly reinterpreted and still preserve the traces of earlier meanings.[4]

In all versions, female protagonists insist on their right to enter sacred areas, despite the ban on women setting foot in certain mountain areas.[5] Moreover, almost all of these women have passed their fertile years. When they cross the marked border and transgress into the forbidden territory, powerful and

---

3 *Sekkyō bushi* 説経節 is the short form of *sekkyō jōruri* 説経浄瑠璃, a rhythmical performance that recalls Buddhist recitation, chanting, and vocal intonations, as in the narration of *Nō* theatre. It became especially popular as a mass entertainment in the early Edo period (Sekiyama 1982, 113–21). All of these genres overlap; it has been argued that the *Nihon ryōiki* 日本霊異記 was the first collection of *sekkyō* (Sekiyama 1992, 56–58). It is certain that the tradition of Buddhist storytelling absorbed many of the local folk stories, and also brought with it continental influences such as Jātaka tales, educational stories of Indian origin. It can even be related to spoken word arts such as *Rakugo* 落語 which is still popular in today's Japan. Its popularity may also be rooted in the entertainment aspect, as the stories should touch and move people to both laugh and cry. It would be adventurous to allege a link between these traditional arts and the *o-warai geinō* お笑い芸能 comedy particularly cultivated in the Kansai region and on TV.

4 For a discussion of orality and literacy in general see Ong (1982); for orality and transmission of *sekkyō* see Matisoff (1986), Sekiyama (1986, 403), Sekiyama (1992, 50–60). *Ishidōmaru Karukaya monogatari* 石堂丸刈萱物語 has also been classified as an example of "written *sekkyō*" as opposed to those stories transmitted orally as lectures (Sekiyama 1973, 309).

5 The prohibition against women entering a sacred area (*nyonin kekkai* 女人結界) applied to many mountain areas in medieval Japan. It still exists in certain regions today, where women are not admitted (*nyonin kinsei* 女人禁制).

threatening forces awake. Among them, the sudden recurrence of menstruation is a common characteristic.[6]

Although the recurrence of menstruation has often been interpreted as a punishment and a heavenly threat, this is far from the only possible interpretation. I argue that the sudden recurrence of menstruation can be seen as a rejuvenating event defying linear measures of time and so accompanying the risky transgression with an empowering act of regaining vital forces. Views of menstruation are themselves heavily conditioned by historical time. While the reason most commonly given for the exclusion of women from sacred areas was ritual pollution (*kegare* 穢れ) because of menstruation and childbirth, there is evidence that in pre-Buddhist Japan women were regarded as life-givers and therefore guardians of fertility (Nakamura 1997, 76; Nishiguchi 1987, 29–31; Shibano 2016, 5–9). Even early Buddhism appears to have regarded the spiritual capacities of all humans, male and female, as equal. Kasahara argues that the idea of karmic disadvantage because of the Five Obstructions[7] and the Three Obediences[8] only started to evolve in Japan around the early 10th century, when a shift towards a patriarchal society started to take place (Kasahara 1975, in Wakita 1992, 61). Before this, blood from menstruation and childbirth was a symbol of vital forces, seen as having healing powers and used in agricultural rites (Suzuki 2005, 218). The social degradation of women associated with their physical functions seems then to be a phenomenon that begins to evolve in the early 10th century and takes shape during the medieval period,[9] a process closely connected to the arrival and spread of Buddhism (Noda 2012, 67).

To elaborate on this proposition, I argue that the local ban on women reflects the patriarchal segregation of women from certain areas of spiritual and political power. The first document mentioning the mountain taboo for women concerns mountain hermit Toran and occurs in the legend collection *The lives of Japanese Immortals* (*Honchō shinsen den* 本朝神仙伝) compiled in 1098 by Ōe no Masafusa 大江匡房 (1041–1111). This timing suggests an interrelation with the spread of Buddhism in medieval Japan. A famous story about the

---

6 As will be shown, the onset of the menstrual flow can either be narrated directly or indicated on a symbolic level.

7 The Five Obstructions (*goshō* 五障) are the five forms of rebirth listed in the Devadatta chapter of the *Lotus sūtra* as prohibited to women.

8 In Confucian thought, women have to subordinate themselves to their fathers, husbands and sons (*sanjū* 三従).

9 After the decline of the aristocracy, a long period of civil wars came to an end when the military government came into power and established its new capital in Kamakura in 1185. The end of the medieval period in Japan is generally equated with the consolidation of the feudal state from 1600 onward.

mother of Kūkai[10] is included in the *Kōbō Daishi Founding Record* (*Kōbō Daishi kaisan engi* 弘法大師開山縁起), which was recorded in 1715 (Hinonishi 1989, 279–80). This story is also embedded in the 17th century sermon tale *Karukaya* 刈萱 and carries remnants of a much earlier oral tradition, itself going back to an independent narrative (Matisoff 2014, 245). This particular story, which was recorded in later collections, may be traced "to before back as early as the twelfth century" (DeWitt 2015, 83), or even earlier when we take the story of Toran into consideration.

In analyzing the use of time-inversion in these stories, I will argue that the onset of menstruation provides a narrative twist that challenges both the measurement of time and the concomitant assignment of a place in society for women. I will briefly touch on chronobiological and diegetic time regarding reproductive matters, and shall then present an analysis of Buddhist legends and folk tales arguing against a focus on female impurity.

2 Physical and Chronobiological Linear Time and Diegetic Time

In contrast to the physical time of growth and decay, diegetic time can be at odds with biological processes. Stories can bend, shortcut, or extend time and thus enable us to see beyond the limited measures and often fragmented perspectives shaped by our physical and social existences. We all know the experience of being in one spot physically while being somewhere else—in the past, in the future—in our minds at the very same moment.

The bending of temporal measures appears frequently in traditional stories concerned with ideas of female fertility. Stories of wondrous children are known all over the world, such as princess Kaguya from the 9th century *Story of the Bamboo Cutter* (*Taketori monogatari* 竹取物語),[11] Tom Thumb, or the Peach Boy, Momotarō. They share common features such as superpowers, extraordinary looks, and a spectacular evolution: a few months are enough to grow from baby to teenager! Another shared characteristic in their family setting is of particular significance here. All of them are received by an elderly couple, an *ojīsan* and *obāsan*, who are clearly described as aged (but bearing their age well). The women have passed menopause and thus are beyond their reproductive years.

10 The monk Kūkai 空海 (774–835) is also known by his posthumous name Kōbō Daishi 弘法大師, or simply referred to as Great Master. He is the founder of the Japanese Shingon Buddhism.

11 For an English translation, see the modern rewriting by Kawabata, translated by Keene (1998). An animated movie adaptation was made by Ghibli studios under the title *The Tale of the Princess Kaguya* in 2013.

The birth of the aged couple's foster children is thus miraculous and does not include sexual intercourse, pregnancy, or childbirth. Princess Kaguya is found in a shining bamboo shoot, and Momotarō is discovered in a giant peach floating down a river. Nevertheless, the couple brings up the child, nursing and caring for the newborn and worrying about its well-being as it makes its way into the world.

Such is the power of stories, and although seemingly far away from our everyday reality, they nonetheless make us consider such up-to-date topics as parenthood, the fragility of the ability to reproduce, and the meaning of relations beyond the post-industrial nuclear family. In their light much of what we consider normal today, can come to seem like invented traditions of shallow origin.

Narratology provides us with categories to label the temporalities of the diegetic world. While temporal composition is based on the order of events, more often than not that temporal order is rearranged in narrative strategies such as flashbacks and other atemporalities. Another feature of storytelling is the duration or manipulation of speed, as in the use of ellipses or simultaneity. Diegetic time also shifts the measure of time by making it possible to repeat events in iterative narration, whether in the narrative or imagined. This frequency then shapes rhythm, and can also express various perspectives, for example when one event is told from the stance of several protagonists.

The stories that I analyze here, of post-menopausal women who experience a sudden reset of their menstrual bleeding contradict a chronobiological measurement of time but make sense in the diegetic world as a narrative rearrangement of the temporal flow. The stories point to female vital powers that became more and more neglected with the spread of Buddhism, as women were assigned a place in society isolated from spiritual, legal, and political influence. I will first discuss the stories of Lady Tamayori 霊拠御前,[12] the mother of monk Kūkai, and then touch briefly on Lady Senri 千里御前, a protagonist who in age and actions contrasts with Tamayori. I will then turn to the older stories of nun Toran 都藍尼, who practiced asceticism herself, and to Shiratōme 白専女, who is known as the mother of Shugendō founder En no

12 The name Tamayori can be found written with the characters 霊拠 or also 玉依. Not only are they pronounced similarly, but the meaning is also very close, as either set of characters could be translated as "Spirit-ascending." This could be yet further evidence for the long time of oral transmission before the stories and its variations were recorded. In other versions, the name of Kūkai's mother is Ato Tamayori あと玉依, sometimes rendered as Lady Akō あこう御前. The family name Akō refers to her descent from Tado-country in Sanuki 讃岐多度群, today's Kagawa province. The suffix *gozen* 御前 is used for persons of high rank and is translated as "Lady" in this paper.

Gyōja.[13] While all these stories show the same basic pattern, they vary in the depiction of the consequences, which may be narrated either explicitly or on a symbolic level. In analyzing their temporal rearrangement of physical processes, I will focus on the awakening of powerful forces and read the stories as carriers of empowerment.

3 Mothers and Wives: Ignore the Rules or Die

Buddhism regularly oscillates between the values of purity and compassion, pondering which of these two is the higher aim. This ambivalence, in combination with traces of previously absorbed beliefs, resonates in the stories about women who enter a forbidden mountain area to search for an ascetic son (McCormick 2012) or for a former husband who is leading a secluded life in training for his spiritual practice. Fathers often are featured only as missing variables. A well-known double version of this story is embedded in the Buddhist sermon tale *Karukaya*. The protagonists of the main story and of the *mis en abyme* respond to the prohibition to enter in opposite ways. Whereas Senri, the protagonist of the story which forms the framework, obeys and does not enter the forbidden area, Tamayori ignores the prohibition and sets foot on the mountain. The two women are thus arranged as contrasting figures, choosing different ways to cope that also lead to different consequences. I shall first briefly describe the frame story of Senri, and then analyze the temporal arrangement of chronobiological linear time in the story of Kūkai's mother Tamayori.

3.1 *The Wife: Lady Senri*

The protagonist of the framing story is Lady Senri, the single mother of Ishidōmaru. His father[14] had left before the baby was born, joined priest Hōnen 法然 (1133–1212)[15] as a mountain ascetic and took on the monk name Karukaya Dōshin 刈萱道心. He moved to the area of Mount Kōya,[16] because this region

---

13 The practice of Shugendō, said to have been founded by En no Gyōja 役行者 (634–ca. 700), combines mountain asceticism and elements of esoteric Buddhism.

14 His original name was Katō Saemon Shigeuji 加藤左衛門重氏, and he was a military commander (*bushō* 武将) in Echizen (located in parts of present-day Fukui and Gifu prefectures).

15 On Hōnen in *setsuwa* legends see Sakaki 1980, 517–26.

16 Mt. Kōya is the spiritual center of the esoteric Buddhist Shingon sect 真言宗. The Kongōbuji 金剛峯寺-temple complex was established in 815 by Kūkai, the founder of Shingon, who had studied esoteric Buddhism in China.

was a restricted area for women, thus ensuring his wife and son could not trace him.[17]

When Ishidōmaru[18] turns 14, he wants to find his father, whom he has missed throughout his childhood. Accompanied by his mother, Senri, who is now 33 years old, he sets out on his quest and travels to Mount Kōya. When the two rest at the Kamuro Inn at the foot of the mountain, Senri learns that she is not allowed to enter the mountain because she is a woman and therefore banned from the area.

To dispel her doubts, the innkeeper tells the story of another woman who disobeyed this rule with terrible consequences. The protagonist of this *mise en abyme* story is not just anyone, but Tamayori, the mother of the revered Kūkai. This certainly adds to the weight of the prohibition and makes clear that absolutely no exception is allowed.

Convinced by this story, Senri remains at the inn and waits for her son's return. Ishidōmaru ascends the holy mountain and continues his search, but returns in vain. Although he has actually encountered his father, Dōshin does not reveal his identity. When he returns to the inn he learns that his mother has died during his absence, and decides to join the mountain monk Dōshin. He does so, but Dōshin takes his secret to the grave. This is the end of a tragic and melodramatic story about a son who lost[19] both his parents—leaving unresolved the deep conflict between filial piety and overcoming mundane attachments. It is a famous story, and various versions and sub-stories of it circulate throughout Japan, both written and orally transmitted. In this paper, I draw

17 His motives for abandoning his family vary depending on the version of the story. In one version, Shigeuji was horrified by the jealousy between his wife and his concubine. One evening he saw their hair turning into fighting snakes, and this experience made him flee from his home and become a monk. In this version, Ishidōmaru is the child of the concubine. The SNKBT version quoted here (Shinoda and Sakaguchi 1999) does not mention female rivalry but says that Shigeuji had a dream about entering the priesthood when his wife was seven and a half months pregnant, and subsequently did so.

18 There is an Ishidō Jizō 石堂地蔵, also known as Karukaya Jizō 刈萱地蔵, in Fukuoka city 福岡市博多区千代3丁目. Other local mentions of Karukaya can be found throughout southwestern Japan, such as a signpost at the bridge crossing before Tofurōmae-station 都府楼前駅 of the Tenjin Omuta line 大牟田線 of Western Railways 西鉄 (Ryū and Sumita 1979, 79), and the Ishidōgawa-river between Hakata and Hakozaki (Ryū and Sumita 1979, 57).

19 The loss of his father can also be interpreted as Ishidōmaru's ignorance due to his attachment to physical and social appearances, and thus to his being deeply rooted in mundane attachments. The same could be said about the loss of his mother, as death is only the end of one's temporal and physical appearance at one stage of the karmic cycle. These thoughts raised by the *Karukaya* story recall the educational Jātaka stories, which lay the ground for further exploration through discussion and learning.

on the version recorded in the 17th century sermon tale *Karukaya*, because it contains the embedded story of Tamayori which is not the case in all other versions.

Senri from the 17th century tale is impressed by the story she is told about the dangers of disobedience and obeys the prohibition. Nevertheless she does not survive—perhaps not despite, but rather because of her obedience to patriarchal laws. Let us now turn to the embedded story of Tamayori and her decision to disobey the mountain ban and set foot on Mount Kōya.

3.2 *The Mother: Lady Tamayori*

The story is recorded in the final episode of the *Kōya* chapter of the sermon tale *Karukaya*, where an innkeeper tells it to explain the ban of women on mountains. It is also recorded in the *Kōbō Daishi Founding Record* (*Kōbō Daishi kaisan engi*). The *mise-en-abyme* story about Lady Tamayori, the mother of revered monk Kūkai, begins with the words "This is the story of the Great Master Kūkai" (Shinoda and Sakaguchi 1999, 288–90). The intention is to reinforce the point that the taboo is to be taken seriously, as it applies even to the mother of a holy man like Kūkai (Miyata 1979, 58–66). The technique of inserting a story within a story also serves to embed an older story and include it in a written text, while, as a means of storytelling, the recurring sequence engages listeners and readers in the loop of the diegetic world precisely by playing with their ordinary sense of the progression of time.

At age 83, Tamayori climbs Mount Kōya to find her son Kūkai, who has retired here for his ascetic practice and meditation, but finds access to the mountain area prohibited to her. Although warned, she insists on continuing her journey into the territory forbidden to women (and horses, and cows by the way). As she crosses the border, the sky becomes cloudy, the earth begins to shake, and a thunderstorm arises, and all of a sudden her menstruation recurs, although she has not menstruated since age 41. Miraculous and wild things happen—and she is reborn finally as Miroku Buddha:

> Well, well, the mother of the Great Master (*daishi no hahago* 大師の母御), she was eighty-three years old, went to meet the Great Master and climbed up Mount Kōya, but all of a sudden the sky became clouded, an earthquake shook the mountain and a thunderstorm came up (*niwaka ni kakikumori yama ga shindo raiden suru* 俄にかき曇り山が震動雷電する).
>
> *SNKBT*, 288; translation my own

The rising of natural forces such as the thunderstorm and earthquake manifest the awakening of cosmic forces, as if they feel her approaching. Tamayori responds to this physically, and starts to menstruate in spite of her old age. The laws of chronobiological time are annulled and the physical time of the body changes from linear to cyclical through that onset of menstruation. Although often interpreted in Buddhist thought as an act of punishment, this episode can also be read as an act of rejuvenation and as a transformation for Tamayori, whose vital forces are re-activated through the act of transgression.

Interestingly, Kūkai hears his mother mention her descent: "I am from Tado-county in Sanuki Province 讃岐多度郡 [...], looking for my child" (SNKBT, 288). As we will see, Shiratōme originates from the same region in today's Kagawa province, which may point to a common origin and to the multilayered versions being orally transmitted and circulating over the centuries throughout Japan.

Kūkai, an obedient son, trusts his mother and through the use of his monk's robe (*kesa* 袈裟) becomes instrumental in her regaining her menstrual flow:

> Kūkai clapped his hands and exclaimed: "It is I, Kōbō, who came here after becoming a priest so long ago! You've done well to climb this far, but the mountain is forbidden to women. All that are male may enter, and all that are female may not. That includes even the birds in the sky and the beasts on the earth." Kūkai's mother was so angry at being excluded from her son's mountain that she twisted up a nearby boulder, which is now called Twisted Rock. Fire rained from the sky (*hi no ame ga orikitareba* 火の雨が降り来たれば), so Kūkai hid his mother in a stony hollow; the boulders there are called the Hiding Rocks. Kūkai's mother wept and spoke: "However great a priest you may be, it is only because of your father's seed and your mother's womb that you're here at all to preach in this latter age [of the dharma]. And instead of inviting your one and only mother to the temple, you're ordering her back down to the village." "It is not because I'm unfilial," Kūkai replied. He took off his seven-paneled stole and spread it on a rock. "Step over this," he said. Because the stole was her son's, the mother was unconcerned. She boldly stepped across, whereupon the blood of her monthly malady (*tsuki no sawari* 月の障) began to fall in tiny drops (*keshitsubu to otsureba* 芥子粒と落つれば). It had ceased when she was forty-one, but now, at the age of eighty-three, it began anew. The stole burst into flames and flew up into the sky (*kaen to natte ten e agaru* 火炎となって天へ上がる).
>
> Translation based on SNKS, 288–89[20]

20 See Kimbrough (2013, 83–84).

This account of the supposed pollution of a holy precinct should explain why women are not allowed to enter the mountain area. At first glance, menstrual blood here appears as a signifier of danger and trespass, as it occurs in a context contrary to the linear and unidirectional body time of the fertile phase in a woman's life. However, once the inversion of measures of time is accepted as a positive rather than negative element, menstrual blood can equally well be seen as a symbol for vital and generative powers (Fukuzaki 2019, 64) in reference to the life-giving vital forces of women in pre-Buddhist time. The dripping blood is said to "drop down like mustard/poppy seeds."[21] Significantly, the smell of mustard seeds was thought to chase away demons, as other examples demonstrate. When, in the *Tale of Prince Genji*, Lady Aoi is threatened in her puerperal period, following childbirth, by a revenging ghost,[22] the monks burn mustard seeds to repulse the demon and protect her. Although it remains unclear what seeds the writer of *Karukaya* had in mind, it seems to be rather a fragrance than an image. It is tempting to be seduced by the bright red of poppy flowers as an equivalent to the color of the blood dripping down, but this should not distract from the fact that Tamayori produces a powerful means to protect herself from evil with her own body.

The contrast between Tamayori and Senri is huge, in both mindset and actions. I argue that these differences showcase a decrease in women's self-confidence in an increasingly patriarchal society, as a shift towards androcentric society increasingly reinforced the dominance of the warrior government of medieval Japan.

The positive implications of Tamayori's story persisted, however, and may be seen in current practice. Today the Jison-in temple in Kudoyama at the foot of Mt. Kōya is a cultic site for women. Female pilgrims offer prayers here for safe pregnancy and childbirth, and for sufficient milk to nurse their babies. Women offer breast shaped talismans in the temple, as described in the 1959 novel *Kii no kawa* 紀伊の川 by Ariyoshi Sawako 有吉佐和子 (1931–1984) and currently on the temple website: "At Jison-in Temple, women can dedicate votive tablets shaped like breasts, a symbol of femininity, to pray for fertility, safe childbirth,

---

21 The characters *keshi* 芥子 today are mostly read as poppy seeds. However, in ancient medical texts, *kaishi* 芥子 refers to mustard seeds, which are a little bit bigger than poppy seeds and blackish red in color. Poppy was brought onto the Japanese archipelago much later, in the Muromachi period (1336–1573). Today, both mustard and poppy seeds are burnt in Buddhist ceremonies. While some obscurity remains, the smell of these burnt seeds has been believed to be an effective demon repellent since ancient times (http://fcmuseum.blogspot.com/2021/04/).

22 The *Genji monogatari* 源氏物語 was written by court lady Murasaki Shikibu 紫式部 (978–ca. 1014) in the 10th century.

successful breastfeeding, and the healing of breast cancer. You can create these votive tablets yourself with your sincerity and heartfelt intentions" (Jison-in 2015). In Southwestern Japan, there are quite a few temples which serve similarly as pilgrimage sites where women can pray for safety and the physical and spiritual stamina to nurse and raise a child, and which also sell talismans and votive tablets with depictions of breasts or women breastfeeding their babies (Iwai 1974, 173–75).

Further, the story of Tamayori points to the stories of Toran and Shiratōme, which have similar characteristics. They are all about elderly women, beyond their reproductive life stage, who enter forbidden territories and awaken threatening powers. The story thus points to another older or alternative version which was orally transmitted over the centuries. When the older story is then included in a written medium, different versions are juxtaposed, creating a further tension through narrative simultaneity. At the same time, the juxtaposition sheds light on the difference between the stories of the 8th century and those set about a thousand years later, each of which bears traces of the mindset of its own time. As will be shown, Tamayori also exhibits some shared characteristics with Toran, such as her brisk and unintimidated attitude, her words of resistance when confronted with the prohibition, and her brave attempt to move on anyway.

4 The Hermit: Nun Toran

Nun Toran is a mountain hermit, many hundred years old, training in asceticism in the Yoshino district in the province of Yamato (today's Nara prefecture). She is the only woman protagonist in the *Honchō Shinsen-den*, and has been described as a "would-be mountain ascetic because she engaged in ascetic practices like the male hermits and ascetics but was prevented from entering their holy places owing to her 'female impurity'" (Kleine and Kohn 1999, 137). This interpretation reflects an often unquestioned assumption that sees the impurity of women as simply a given fact, rather than as a belief that may have evolved over time. Indeed, Toran does not succeed in entering a prohibited area, but the fact that she attempts to do so suggests that in itself the possibility of entering the holy area. In one version of the stories, Toran questions the prohibition and refuses to accept it with words very similar to those of Tamayori. She also declares herself to be not an ordinary woman, but someone with the training and experience to ascend into the sacred area. But when Toran wants to enter the mountain, thunderstorms arise and something terrible happens to her. Nothing remains of her except the traces of her fingernails, her walking staff and a new spring (Suzuki 2005, 82). In another version of the story, Toran chants a spell and ascends the mountain with a dragon.

Similar stories of the taboo feature a huge white snake,[23] which in its frequency in this context can be related to the female figure banned from the mountain area. As a symbol, the snake contains a variety of overlapping imagery, such as the rain and thunderstorm deity, as well as temporal measures in the cyclical renewal of the body through molting, as seen in sacred snake deities in local beliefs such as Ōmononushi 大物主神 on Mount Miwa.

Toran is also the first figure to be depicted as provoking the anger of the mountain deity by trespassing on the liminal area (*kyōkai* 境界) and climbing the holy mountain (*tōhai* 登拝)[24] of Mount Kinpu[25] in Yoshino district. The story goes that the mountain area becomes stormy, and heavy rain and landslides occur:

> Nun Toran was a person of the country Yamato. She followed the law of Buddha and earned a long life. How many hundreds of years [she lived] is not known. She lived at the foot of Mount Yoshino, and devoted herself to asceticism day and night. She set off to visit the forbidden Mount Kinpu (*Kinpu-san ni kinji agaran to hossu* 欲攀上金峰山). Thunder and lightning, a violent thunderstorm came up, and she could not get there (*raiden hekireki shi, tsui ni itaru o ezu* 雷電霹靂、遂不得到). The soil of this mountain was laid out with gold. Because it awaits the advent of Maitreya [Miroku bodhisattva aka Jison], the deity Kongō Zaō [Vajrasattva] protects it. This is a land of the precepts, and for this reason women may not access it (*noynin o tōzezaru* 不通女人). The place where her walking stick [fell] transformed into a tree, and the place where she got stuck collapsed and became a fountain. The marks of her fingernails still exist.
>
> *Honchō shinsen den*, translation my own

23 Snakes and dragons are associated with water and rain in Asian mythologies. In Japanese, the flash in a thunderstorm is called *inazuma* 稲妻, literally the spouse of the rice plant. which hints at the importance of rainfall for an agricultural society. The depiction of an electrical storm in the shape of a dragon is frequent.

24 In 2016, August 11 became a national holiday in Japan called "Mountain Day."

25 Today known as Mount Ōmine (also, but less famously, Mount Sanjō 大峰山山上ヶ岳), it is one of the earliest mountains to have been opened to Shugendō. The legend goes that En no Gyōja introduced here the deity Zaō Gongen, 蔵王権現, a central and powerful deity of Shugendō religion, also known as Kongō Zaō Bodhisattva 金剛蔵王菩薩 (Vajraattva) (Suzuki and Blair 2009, 142). The deity is associated with Buddha Sakyamuni and the Bodhisattva Maitreya as a manifestation according to the syncretistic *honji suijaku* theory.

In another version of the story, however, Toran argues with the guardian deity, and displays her supernatural powers, only hinted at above:

> "Although I have a female body, my discipline is pure and I have numinous experiences. How can you think that I am an ordinary woman?" So she climbed up Mitake. Suddenly there was thunder and lightning; it became dark, and she lost her way. She abandoned the staff she was holding. This staff spontaneously grew thicker and gradually became a big tree. Thereupon Toran chanted a spell, mounted a dragon, and ascended the mountain on its back. When they arrived at the fountain, however, they could not proceed any further. In anger, Toran stamped her foot so that the rocks on the mountain peak broke into pieces. The pond of the dragon is [therefore] covered by rocks. Two traces [of Toran] still exist today. People say that she attained the way of longevity, and nobody knows where she is.
>
> BUSSHO KANKŌKAI 1984, 346a; translation Kleine and Kohn 1999, 155

The ability to create a fountain, grow a tree, split mountains and ride a dragon serve as proof of the highly evolved powers and spiritual abilities of Toran. These can be read as references to the existence of female *miko* 巫女, mediums and shamans practicing in areas not allowed to ordinary people. There is a parallel to the non-accessible sacred areas in the mountains (*kannabi* 神奈) of ancient times and to the Buddhist areas later made taboo (*shichiri kekkai* 七里結界), both of which were prohibited to most humans, regardless of their sex.[26] The gradual shift from the general taboo (*kinki* 禁忌) to the exclusion of women in particular occurred over time, justified by the idea that menstruation and childbirth caused female impurity (Suzuki 2005, 135, 144). These characteristics are also seen in stories of female Daoist practitioners in continental collections (Despeux 1990, Cahill 1990).

This overlapping and ambivalent imagery can also be found in the story of Shiratōme, the mother of the legendary founder of Shugendō mountain asceticism, En no Gyōja. A wooden statue of Shiratōme dated from the 14th century is revered as an object of worship in the Sakuramotobō temple in Yoshino. The

26 Prohibited venues were dwellings of deities (*iwakura* 磐座), to which it was believed that deities or spirits (*tama*) were attached (*yoru*). As these could be rocks or trees, in pre-Buddhist practices the spirits could also come to dwell temporarily in a human being, which was the role of the shaman maidens (*miko* 巫女). Female hermits such as these seem to have originally practiced in the mountains along with men, but were ousted by the increasing number of male mountain ascetics (Suzuki 2005, 135ff). The name of Tamayori could be further evidence of this, see footnote 12.

accompanying text reveals her relation to Toran clearly: "En no Gyōja's mother is called by the names Princess Totoki 渡都岐比売, Shiratōme, or Torame 刀自女" (Sakuramotobō 2004–2022). She is said to have encountered a large snake when she was about to transgress into the forbidden area, as illustrated by an inscription in the Yoshino region:

> About the ban on women (*nyonin kinsei* 女人禁制) on Mount Ōmine, there is a story. The mother of En no Gyōja, Shiratōme, was worried about her son who was practicing asceticism. When she came to the snake valley of Dorogawa-river, her way was cut off by a large snake (*daija*[27] *ni yukute o habamare* 大蛇に行く手を阻まれ), and she remained at the Dorogawa-village.
>
> VILL TENKAWA NARA, n.d.; translation my own

The influence of time as a measure of meaning appears here as well, as over centuries of oral transmission, these figures seem to have overlapped and merged with the protagonists of later ages, such as Tamayori. Including these last female protagonists also reveals a gradual shift towards female obedience and subordination to patriarchal structures. Toran was training as a mountain hermit, Tamayori in fury refuses to accept the mountain ban for women, while Senri shows no resistance at all and dies. Parallel to this, the appropriation or inclusion of these figures into Buddhism can be observed as well. While Toran remains relatively unappropriated, she can be associated with esoteric mountain asceticism through the connection to Shiratōme. In contrast, Tamayori is syncretistically associated with a figure from the Buddhist pantheon. Many centuries later, the location of Senri's death was transformed into a Buddhist pilgrimage site for women in relation to fertility and thus the continuation of life through the female body.

5 Conclusion

The stories presented here reach back to the 11th century and even further. Their shared features, such as the common origin of Tamayori and Toran in Sanuki Province, along with the continental influences, further point to an oral transmission, and with this the development of multiple variations and layers to the story.

27 The characters 大蛇 can alternatively be transcribed as *orochi*, which semantically refers to the mythological monster serpent.

In all these stories, fertility issues are addressed directly (through menstruation) or indicated on a symbolic level (through snake imagery). The historical conditioning of this view appears in the gradual shift from fertility worship (*hōjō* 豊穣) to the exclusion of women, as seen in the first mentions of blood pollution (*ketsue* 血穢) in the early 10th century. The gradual exclusion of women from spiritual, political, and social participation is mirrored in the shift from a general taboo towards an increased belief in female impurity. Fertility as a physical phenomenon becomes connected to impurity and so also to the exclusion of women from higher religious levels. A linear and misogynistic way of measuring time, which aligns life stages with one's place in society, is challenged through the diegetic counter-measurement of time. Often interpreted as punishment for female transgression, the stories discussed in this paper rather reveal an underlying access to physical forces that can be traced back to pre-Buddhist times. In so doing they also reveal an area of conflict between patriarchal law and the subliminal motives of vital force. Compassion and purity no longer are matters of the mind, but of the human—especially the female—body. The necessity of retiring from everyday societal life, as in the time needed to recover after childbirth, was handled as a taboo of "not being seen" and transformed into a ritual exclusion of women (Matsumoto 2020, 73). Simultaneously, the female womb was appropriated in a spiritual way, as seen in the "becoming" of Jison and the subsequent incorporation of Tamayori into the Buddhist pantheon.[28]

The stories discussed in the paper give insight into a cyclical flow of time, where it is possible to recover vital forces and connect with spiritual powers beyond the linear measurement of time in chronobiology. In the narrative rearrangement of the temporal order, the unidirectional patriarchal measure of time is suspended and makes place for subliminal powers awakened by the entering of sacred sites.

Menstruation has been viewed as a reason for the exclusion of women from spiritual areas. At the same time, the aspect of fertility and generative powers associated with the female body have been successfully incorporated into the Buddhist canon. The temporal twists in the diegetic world of the stories presented reflect the re-integration of pre-Buddhist folk beliefs of fertility into Buddhist thought more generally.

28 The story of Toran also shows an association with the dual worlds of esoteric Buddhism, consisting of the dual worlds of the Womb Realm (also Matrix Realm *taizōkai* 胎蔵界) and the Diamond Realm (*kongōkai* 金剛界).

References

Primary Sources

Shinoda, Jun'ichi 信多純一, Sakaguchi, Hiroyuki 坂口弘之 (eds.). 1999. *Shin Nihon Koten Bungaku Taikei* (SNKBT) *90: Kojōruri sekkyōshū* 新日本古典文学体系 90: 古浄瑠璃説教集. Iwanami shoten: 247–314.

Yatarō, Muroki 室木弥太郎 (ed.). 1977. *Shinchō Nihon Koten Shūsei* (SNKS) *8: Sekkyō-shū* 新潮日本古典集成 8: 説教集. Shinchōsha 1977: 11–77.

Keller Kimbrough, R. 2013. *Wondrous Brutal Fictions: Eight Buddhist Tales from the Early Japanese Puppet Theater*. Columbia University Press.

Honchō shinsen den 本朝神仙伝: *National Archives of Japan Digital Archive* (NAJDA) 国立公文書館デジタルアーカイブ. Accessed April 19, 2025. https://www.digital.archives.go.jp/item/735622.

Busshō Kanko kai 佛書刊行会. 1984. Shingon fuhō-den 真言付法傳. In Dainihon bukkyō zensho 大日本仏教全書 106. Tokyo: Meichō fukyū kai.

Kleine, Christoph and Kohn, Livia. 1999. "Accounts of Japanese Immortals. Translation of the *Honchō shinsen-den*." In Kleine and Kohn 1999, 143–84.

Secondary Sources

Araki, Shigeru 荒木繁. 1998. "Sekkyō no seisui" 説教の盛衰. In: *Jōruri no tanjō to ko-jōruri. Iwanami kōza Kabuki Bunraku 7* 浄瑠璃の誕生と古浄瑠璃。岩波講座歌舞伎・文楽　第七巻. Iwanami shoten: 71–93.

Bonaventura, Ruperti (ed.). 2019. *The Body in the Japanese Performing Arts. Death and Life, Puppets and Artificial Bodies* 日本の舞台芸術における身体. Kyoto: Kōyō shobō.

Cahill, Suzanne. 1990. "Practice Makes Perfect: Paths to Transcendence for Women in Medieval China." *Taoist Resources* 2, no. 2: 23–42.

Deal, William E. 2006. *Handbook to Life in Medieval and Early Modern Japan*. Facts on File.

Deal, William E. 2017. "Religion in archaic Japan." In *Routledge Handbook of Premodern Japanese History*, edited by Karl F. Friday, 187–201. Routledge.

DeWitt, Lindsey Elizabeth. 2015. *A Mountain Set Apart: Female Exclusion, Buddhism, and Tradition at Modern Ōminesan, Japan*. University of California. Accessed July 2, 2024. https://escholarship.org/uc/item/24q896hq.

Despeux, Catherine. 1990. *Immortelles de la Chine anciénne. Taoïsme et alchimie féminine*. Pardès.

Dix, Monika. 2009. "Saint or Serpent? Engendering the Female Body in Medieval Japanese Buddhist Narratives." In *The Body in Asia*, edited by Brian S. Turner and Zheng Yangwen, 43–58. Berghahn Books.

Faure, Bernard. 2003. *The Power of Denial: Buddhism, Purity, and Gender*. Princeton University Press.

Fukuzaki, Takao 福崎孝雄. 2019. "Chi ni taisuru kegare ishiki" 血に対するケガレ意識. In *Gendai mikkyō* 現代密教 8: 62–74.

Hinonishi, Shinjō 日野西眞定. 1989. "Kōya-san fumoto Karukaya-dō no hassei to kinō—toku ni Senri gozen no mikoteki seikaku ni tsuite" 高野山麓刈萱堂の発生と機能—特に千里御前の巫女的性格について—. In *Josei to bukkyō 4. Miko to megami* 女性と仏教 4 巫女と女神, edited by Kazuo Ōsumi 大隅和雄 and Nishiguchi Junko 西口順子, 241–90. Heibonsha.

Hinonishi, Shinjō 日野西眞定 (ed.). 1988. *Kōbō Daishi shinkō* 弘法大師信仰. Yuzankaku shuppan.

Histrip. 2017. "Hisomerareta rekishi no hōko. Tenkawa-mura ni imo mo nokoru Shugendō no densetsu ni semaru rekishitabi" 秘められた歴史の宝庫天川村に今も残る修験道の伝説に迫る歴史旅. Accessed April 18, 2025. https://www.histrip.jp/171104nara-yoshinogun-1/#i-2l.

Hobsbawm, Eric and Terence Ranger (eds.). 1983. *The Invention of Tradition*. Cambridge University Press.

Hisano, Toshihiko 久野俊彦. 2009. *Etoki to engi no fōkuroa* 絵解きと縁起のフォークロア. Shinwasha.

Ihara, Kesao 井原今朝男. 2013. *Shijitsu Chūsei bukkyō 2. Sōsō monoimi to jiin kinyū / shinbutsu kōsō no jitsuzō* 史実中世仏教 2. 葬送物忌と寺院金融・神仏抗争の実像 2. Kohzansha.

Ijūin, Yōko and Sachiko Kawai. 20 "Gender and family in the archaic and classical ages." In *Routledge Handbook of Premodern Japanese History*, edited by Karl F. Friday, 202–15. Routledge.

Ishii, Nobuko. 1989. "Sekkyō-bushi." *Monumenta Nipponica* 44: 283–307.

Iwai, Hiromi 岩井宏美. 1974. *Ema. Mono to hito no bunkashi* 12 絵馬. ものと人の文化史 12. Hōsei University Press.

Jackson, Reginald. 2019. "Desiring Spectacular Discipline." *Asian Theatre Journal* 36, no. 1: 49–78.

Jison-in 慈尊院. 2015. Accessed April 18, 2025. http://jison-in.org/about/top.html.

Kasahara, Kazuo 笠原一男. 1975. *Nyonin ōshu shisō to keifu* 女人往生思想と系譜. Yoshikawa kōbunkan.

Kawabata, Yasunari. 1998. *The Tale of the Bamboo Cutter*, translated by Donald Keene. Kodansha International.

Keller, Kimbrough, R. 2008. *Preachers, Poets, Women, and the Way: Izumi Shikibu and the Buddhist Literature of Medieval Japan*. Center for Japanese Studies.

Keller, Kimbrough R. 2013. *Wondrous Brutal Fictions: Eight Buddhist Tales from the Early Japanese Puppet Theater*. Columbia University Press.

Kikuchi, Hitoshi 菊地仁. 1978. "Sekkyō Karukaya to Kōyamaki" 説教刈萱と高野巻. In *Denshō bungaku kenkyū* 伝承文学研究 21.

Kitagawa, Joseph Mitsuo. 2021. *On Understanding Japanese Religion.* Princeton University Press.

Kleine, Christoph and Kohn, Livia. 1999. "Daoist Immortality and Buddhist Holiness: A Study and Translation of the honcho shinsen-den." *Japanese Religions* 24, no. 2: 119–96.

Kobayashi, Kenji 小林健二. 1995. "Katarimono no tenkai (2)—sekkyō *Karukaya* to *Kōya no maki*" 語り物の展開 (2)ー 説教「刈萱」と「高野の巻」. In *Sanbun bungaku "Monogatari" no sekai* 散文文学〈物語〉の世界 3, no. 23.

Matisoff, Susan. 2014. "Nine Holy Horrors: The Sermon-Ballads of Medieval and Early Modern Japan." In *Flowing Traces: Buddhism in the Literary and Visual Arts of Japan*, edited by James H. Sanford, William R. LaFleur and Masatoshi Nagatomi, 234–62. Princeton University Press.

Matisoff, Susan. 1986. "Sekkyō texts and the concept of 'oral literature'" 説教正本といわゆる口承文学. *Kokusai Nihon bungaku kenkyūshū-kai kaigiroku* 国際日本文学研究集会会議録 9: 23–32.

Matsumoto, Ikuyo 松本郁代. 2020. "Nihon shūkyō ni okeru 'sei' no tayōsei" 日本宗教における「性」の多様性. In *Nihon shūkyō no shinkōkai* 日本宗教の信仰界, edited by Fumiko Satō 左藤文子 and Satoshi Ito 聡井藤, 69–91. Yoshikawa Kōbunkan.

McCornick, Melissa. 2012. "Mountains, Magic, and Mothers: Envisioning the Female Ascetic in a Medieval Chigo Tale." In *Crossing the Sea: Essays on East Asian Art in Honor of Professor Yoshiaki Shimizu*, edited by Gregory P. A. Levine, Andrew M. Watsky, and Gennifer Weisenfeld, 107–33. Princeton University Press.

Miyata, Noboru 宮田登. 1979. *Kami no minzokushi* 神の民族史. Iwanami shoten.

Miyata, Noboru 宮田登. 1999. *Kegare—sabetsu shisō no shinsō* ケガレ 差別思想の深層. Kaihō shuppansha.

Morrison, Barbara S. 2006. *Body Rhetoric: Women en Route to Salvation in Oguri and Pilgrim's Progress*. University of North Dakota.

Muroki, Yatarō 室木弥太郎 (ed.). 1977. "Kaisetsu" 解説. *Shin Nihon Koten Shūsei. Sekkyōshū* 新日本古典集成 説教集, 393–423. Shinchōsha.

Nakamura, Kyoko Motomachi. 1997. *Miraculous Stories from the Japanese Buddhist Tradition: The Nihon Ryoiki of the Monk Kyokai*. Harvard University Press.

Nakano, Yuko. 1988. "Women and Buddhism—Blood Impurity and Motherhood." In *Women and Religion in Japan*, edited by Akiko Okuda and Haruko Okano, 65–85. Harrassowitz Verlag.

Nishiguchi, Junko 西口順子. 1987. *Onna no chikara: kodai no josei to bukkyō* 女の力ー古代の女性と仏教. Heibonsha.

Nishiguchi, Junko 西口順子. 2006. *Chūsei no josei to bukkyō* 中世の女性と仏教. Hōzōkan.

Noda, Renate. 2012. *Reisende Frauen aus der Edo-Zeit und ihre Reisetagebücher*. TenDenZen.

Okano, Haruko. 1991. "Die Stellung der Frau in der japanischen Religionsgeschichte." In *Japan. Ein Land der Frauen?*, edited by Elisabeth Gössmann, 34–55. Iudicium.

Ong, Walter. 1982. *Orality and Literacy*. Methuen and Co.

Ōsumi, Kazuo 大隅和雄. 1983. "Josei to bukkyō—kōzō to sono haha" 女性と仏教　高増とその母. In *Shiron* 1983: 36.

Ruch, Barbara. 2002. *Engendering Faith: Women and Buddhism in Premodern Japan*. Center for Japanese Studies: The University of Michigan.

Ryū, Kankichi 劉寒吉 and Sumita, Yoshihisa 角田嘉久. 1979. *Fukuoka no densetsu* 福岡の伝説. Kadokawa shoten.

Sakaguchi, Hiroyuki 坂口弘之. 1992. "Kōya no denshō nidai—'Kōbō Daishi godenki' 'Uguisu no Mida no koto'" 高野の伝承二題ー「弘法大師御伝記」「鶯の弥蛇のこと. *Studies in the humanities: Bulletin of the Graduate School of Literature and Human Sciences* 44, no. 13: 1069–83.

Sakaki, Taijun. 1980. 榊泰純. *Nihon bukkyō geinō-shi kenkyū* 日本仏教芸能史研究. Kazama shoten.

Sakuramotobō 櫻本坊. 2004–2022. "En no Gyōja bokō zō" 役行者母公像. Accessed April 19, 2025. https://sakuramotobou.or.jp/about/ennogyojas-mother.html.

Sato, Eiko. 1989. *Die Frau im alten Japan*. Brücken Verlag.

Satō, Fumiko 左藤文子 and Ito, Satoshi 井藤聡 (eds.). 2020. *Nihon shūkyō no shinkōkai* 日本宗の信仰界. Yoshikawa Kōbunkan.

Sekiyama, Kazuo 関山和夫. 1973. *Sekkyō no rekishiteki kenkyū* 説教の歴史的研究. Kyoto: Hōzōkan.

Sekiyama, Kazuo 関山和夫. 1982. *Bukkyō to minkan geinō* 仏教と民間芸能. Hakusui sha.

Sekiyama, Kazuo 関山和夫. 1987. *Sekkyōshū* 説教集. Chūō kōron.

Sekiyama, Kazuo 関山和夫. 1992. *Sekkyō no rekishi. Bukkyō to wagei* 説教の歴史　仏教と話芸. Hakusui U Books.

Sekiyama Kazuo Hakushi Kiju Kinen Ronshū Kankōkai 関山和夫博士喜寿記念論集刊行会 (eds.). 2006. *Bukkyō bungaku geinō: Sekiyama Hakushi kiju kinen ronshū* 仏教文学芸能：関山博士喜寿記念論集信多純一. Sekiyama Kazuo Hakushi Kiju Kinen Ronshū Kankōkai.

Shibano, Noriko. 2016. *Transformation d'un Tabou: La Menstruation dans la Société Japonaise*. Disseration. Montréal: Université du Quebec.

Shioiri, Hidetoshi 塩入秀敏. 2001. "Migawari Jizō densetsu-kō—kikyū kinan-gata migawari Jizō, toku ni Kanayaki-Jizō densetsu ni tsuite" 身代わり地蔵伝説小考ー危急危難型身代わり地蔵伝説、とくに金焼地蔵伝説について. In *Ueda Joshi Tanki Daigaku kiyō* 上田女子短期大学紀要 25: 21–32.

Suzuki, Masataka 鈴木正崇. 2005. *Nyonin kinsei* 女人禁制. Yoshikawa Kōbunkan.

Suzuki, Masataka 鈴木正崇. 2015. *Sangaku shinkō—nihon bunka no kongen o saguru* 山岳信仰一日本文化の根源を探る. Chūō kōron.

Suzuki, Masataka 鈴木正崇. 2022. "Exclusion of Women and Mountain Beliefs" 女人禁制と山岳信仰. *Tetsugaku collection* 哲学集 149: 145–79.

Suzuki, Shōei and Heather Blair. 2009. "The Development of Suijaku Stories about Zaō Gongen." *Cahiers d'Extrême-Asie* 18: 141–68.

Takeda, Kazuaki 武田和昭. 2017. "'Karukaya' no monogatari—'Kōya-maki' to Shikoku no Kōbo Daishi denshō to no kankei" 「かるかや」の物語一「高野巻」と四国の弘法大師伝承との関係. In *Sekkyō. Hito wa shinbutsu ni nani o takusō to suru no ka* 説教人は神仏に何を託そうとするのか, edited by Kōbe joshi daigaku koten geinō kenkyū sentā 神戸女子大学古典芸能研究センター, 103–18. Ōsaka: Izumi shoin.

Tokuda, Kazuo. 1976. "Sekkyō toki to shoki sekkyō bushi no kōzō." *Kokubungaku kenkyū shiryōkan kiyō* 2 (March): 19–46.

Ushiyama, Yoshiyuki. 2009. "The historical development of the exclusion of women from the sacred places (*nyonin kinzei*) in Japan." *Acta Asiatica* 97: 39–55.

Vill Tenkawa Nara. N.d. "修験道体験奈良県天川村大峯山." Accessed April, 18 2025. https://www.vill.tenkawa.nara.jp/tenkawa/ds.

Wakita, Haruko 脇田晴子. 1992. *Nihon chūsei josei-shi no kenkyū. Sabetsu yakuwari buntan to bosei, kasei, seiai* 日本中世女性史の研究 差別役割分担と母性・家政・性愛. Tokyo Daigaku shuppankai.

Sawada, Tsunayoshi 達澤田進. 2023. "Shiratōme" 白専女. Accessed June 27, 2023. https://yamabusi.gozaru.jp/siratoume.html.

Yamaguchi, Atsushi 山口敦史. 2003. "*Nihon ryōiki* no joseikan. Densetsu no hyōgen o ika ni yomuka" 『日本霊異記』の女性観 伝説の表現をいかに読むか. *Nihon bungaku* 52, no. 1. Accessed June 24, 2023. https://doi.org/10.20620/nihonbungaku.52.1_1.

Yano Chiyomi 矢野治世美役行者母公像. 2019. *Kōya-san to hi-sabetsu min: "Kongō buji hinamiki" no sekai* 金剛峯寺日並記」の世界. Ōsaka: Kaihō shuppansha.

CHAPTER 10

Narrative Time in Kazakh *Sybyzghy* Performance

Xiaoshi Wei

Abstract

Sybyzghy, a bi-phonic type of flute, is a crucial element in the Kazakh gatherings of Xinjiang and Inner Asia. Each *sybyzghy* tune relates to a unique folktale, of ancient warriors, modern rebel heroes, animals and hunters, rivers and their waves, and famous orator-poets of the Kazakhs, with stories drawn from the 10th through the 20th century. This chapter considers how time, in these historically informed performances, is malleable and multifaceted, depending on people's ideas about their history, nation, and ethnic clan. Examining the indexicality of the pieces, I analyze their motifs and narrative rhetoric, showing how they constitute temporal framings for historical times. These temporal aspects link present-day performance to living history, forming a crucial element in the cultural identity of the Kazakh people, while also enabling Kazakh communities to use this shared imagery of historical figures, perennial narratives, and symbolic meanings to affirm their diverse identities.

Keywords

bi-phonic flute – Kazakh – Inner Asia – Central Asia – Xinjiang – sonic metaphor – performativity of time – musical narratives

This article explores the experience of time and the way it is measured in *sybyzygy* performance in Xinjiang (northwestern China). Specifically, I will be focusing on the way understandings of temporality help to link multiple visions of Kazakh cultural history to contemporary meanings among their people. *Sybyzghy* [pronounced *Suh-buh-z-guh*] is a bi-phonic end-blown flute with origins in the Altaic Mountains region of Western Mongolia. Historically and up to the current day, family gatherings, clan assemblies, and celebrational parties remain an important part of collective Kazakh life. Playing the flute is a key accompaniment to narrative storytelling at such events, helping to place recited stories in a balanced, harmonious, and soothing context. In this way,

 | DOI:10.1163/9789004723634_012

the *sybyzghy* assists people in narrating folktales invoking historical figures, real and imagined landscapes, and social commentary. *Sybyzghy* performances mix verbal content and sounds in order to guide and "perform" the sociality of a gathering by echoing various visions drawn from Kazakh history and cultural memory. In these narratives, performers and audiences collectively operate across three types of performative 'time'—folkloric, mythological, and historical—creating sonic indices, that is to say, links, between particular topics and specific audible and temporal qualities. The result is a shared atmosphere in which the current generation of Kazakhs can engage with particular historical imaginaries, joining the two through a common measure.

As an Inner Asian bi-phonic instrument, the sound of *sybyzghy* has two parts—the droning part, rough and hoarse, and the higher part, delicate and melodious, similar to the throat singing familiar to the Western audience. The performative and timbral qualities of *sybyzghy* have been discussed by various ethnomusicologists (Pegg 1991, 2000, 2021, 2024; Post 2024; Levin 2006; Wei 2024), who have reflected on its construction, cultural roles, and sonic quality within the Inner Asian landscape.[1] In one earlier, closely related article (Wei 2024), I explore the way *sybyzghy* recordings are used by Kazakhs in northern Xinjiang[2] to affirm identities shaped around nationhood, regionality, and clans, discussing the power of recordings to signify a symbolic vision of collective memory. This current article extends my research, looking in more detail at the temporal aspects of *sybyzghy* performance which link present-day performance to a living history in the cultural imaginary of the Kazakh people. It is through this shared imagery of historical figures, perennial narratives, and symbolic meanings that Kazakh communities affirm their diverse identities. As I found out during my fieldwork between 2013 and 2019, it is important for various Kazakh families to perform stories, such as those of Genghis Khan (12th century), Qorqyt (8th–10th century), Asan Qayğı (14th century), and Beysenbi (19th century), not as a direct recounting of history, but through a more personal, mythologized retelling in which time is not necessarily linear.

1 The primary *sybyzygy* performance settings that this article addresses are gatherings in villages as opposed to performances on professional stages, the latter of which are more readily accessible to a general audience. In collective village gatherings, *sybyzghy* playing is not treated as professional artistry, but as a channel to communicate emotions, feelings, and historical visions, and the events that feature its playing are a way to organize villagers and clan members to get together and discuss village affairs.

2 Xinjiang, officially the Xinjiang Uyghur Autonomous Region, is the largest province of modern-day China, directly bordering Mongolia to the east and Kazakhstan to the northwest. It is the established residential region for several non-Han Chinese groups, including but not confined to Uyghurs, Kazakhs, Kirghizs, and Mongolians.

Equally, it is also essential for them to perform such stories in a collective setting, as time must be specifically dedicated to these performances to afford them their proper significance.

I devote this current article to expanding research questions on time, historical visions, and the cultural understanding of temporality that leads to these visions: which specific historical eras and their interpretations relate to ideas of Kazakh nationhood? What mechanism of temporality helps people construct the imaginaries of such times and visions? I am guided in this by Becker and Becker's (1981) idea of the iconic power of music—the "coincidence" of music and life in the "workings of time" (207). I also draw from literature on music in narratives (Feld 1984), and musical semiosis, the creation and interpretation of signs and symbols (Turino 1999). Through this framework, I will reflect on the temporality of *sybyzghy* performance, explaining how its sounds and narratives index or create links with visions of the historical past.

1 Concept of Time in Kazakh Verbal Performances

This article forms a dialogue with studies such as Janabekova 2012, which examine the interwoven nature of space and time as the "fundamental elements of Kazakh world image" (Samuratova, Musalimov, Albytova, Makhanov, and Omarbekova 2015), projected cycles of time in temporally situated Kazakh idioms (Abdramanova S. 2019), and the deeply rooted metaphors of time in the Kazakh language (Beknazarova 2022). Together, this research makes it clear that Kazakh temporality, the cultural understanding of time, not only centers on the play of language, but also relates to ancestral images, the vast landscape of the region's steppe environment, and the lineage of expressive forms, such as music and oral performance.

From a broad, state-level perspective, ethnomusicologist Margarethe Adams (2020), in her monograph, *Steppe Dreams: Time, Mediation, and Post-socialist Celebrations in Kazakhstan*, examines "the political and public aspects of temporality" (5) and "the political significance of temporality" (3) in cultural presentations in Kazakhstan, investigating also the "personal, interior, and emotional aspects of the way time is experienced" (5). Examining both intangible and tangible cultural artefacts, she writes: "Like dreams, such images and music continually refract and echo, carrying temporal meanings about nationhood, about our possible or promised futures" (3). In observing the plurality of past *time*s created through the perception of *place*s, she claims that the significance and meanings of these can then be accumulated in the present (7). Although operating at a more intimate scale, *sybyzghy* performance occurs within a similar cultural understanding of temporality, acting as a way to crystalize a cultural understanding of time through performance, and consequently contributing to the establishment of sonic identity for participants.

Within this field of research, Abdramanova's (2019) work is particularly related to my study of time inside *sybyzghy* performances, especially the assertion that the Kazakh concept of time is "highly lexicalized," with "certain words that define a definite age of the human beings assigned to them" (5). As I will show in the *Indexicality* section of this article, such a "lexicalized" concept of time can be found in the indexical (symbolic) system, indicating not only that perceptions of time depend on verbal performance, but also that there is a system of linguistic understanding in which each verbal symbol uniquely defines its temporal quality, conveying specific meanings and a particular sentiment of historical time to the participants in a performance. Abdramanova (2019) further uses "moving time" and "moving ego" to link the feelings of selfhood and nationhood, through the various eras implied within verbal performance. Relating to this idea, I will show, in my section, "Historical Figures, Imaginaries, and Their Times," the plurality of ideas from which Kazakh ancestral ideas are formed through the narratives of diverse times and their heroic figures.

2 Indexicality of *Sybyzghy* Performance

In the culture of *sybyzghy* performance, music and language are fundamentally interwoven. *Sybyzghy* pieces, mostly comprised of 30-second singular cycles, lack independent melodic structures that could assist the performers in telling a story, and must be performed within a verbal narrative context. A complex weave of music and language is thus offered up for interpretation. Although the propositional text accompanying the music can be analyzed using a linguistic approach, the focus of this article is rather on the emotive 'pulses'[3] that are created from and eventually go beyond language. *Sybyzghy* artists create a unique way of organizing language that links the sound of *sybyzghy* with personalized accounts. Performers not only express explicit things, but also attach emotive meanings to them; the *sybyzghy* performance is not only structurally linguistic, but also features a system of non-verbal signs that express meaning.

Sybyzghy, although in past times occasionally played as an instrument for collective entertainment, more often functions as an introduction to conversation: an instrument that creates impressions of nature, animals, and events that echo the past of Kazakh clans. Gatherings for decision-making and other

3 In this context, "pulse" can be understood in Becker and Becker's (1981) language through its "coincidence with life" (205).

discursive gatherings fuse narrative and *sybyzghy* performance, applying musically affective 'signs' to cue the expected behaviour of conversation participants. For example, introductory parts of a piece can serve to indicate the intended mood of an upcoming conversation, while a performer's verbal utterances between pieces may indicate how detailed the musical 'conversation' itself may become. Meanwhile, the temporal quality of a piece may serve as an index, a symbolic indicator, of the judgement and decision to be made in the upcoming verbal conversation. In Turino's (1999) terms, indexical communication, that is, communication implicit in context, is "most prominent in intimate groups such as married couples, families, close friends, and further down the continuum, in small close-knit communities or neighbourhoods" (235). As groups linked through tribal, clan, and village ties, small Kazakh clans fall into such a type of group communication. Played among groups of Kazakhs that deeply share cultural markers of experience and emotion, *sybyzghy* pieces together with their verbal narrations serve as an indexical system that makes up a critical scaffold in Kazakh communication, registering a shared knowledge of symbolic tropes and their implicit meaning or point of reference.

Above all, the sound of *sybyzghy* provides an acoustic basis for setting the mood of a gathering and leading the audience into a shared emotive state. The acoustic basis is primarily reflected in the sonic dialectic between *qoñır* (soft and pacifying) and *aştı* (dynamic and 'spicy') pieces. At the beginning of a performance, the local leader usually plays a *qoñır* piece to set the atmosphere for the whole event. Performers then organize proceedings through the different topics of the *sybyzghy* pieces, even while improvising. Each topic has its unique acoustic quality, melodies and timbre, like the acoustic representation of *aştı* and *qoñır* pieces. Spoken conversation is propelled into a particular direction by the musical pieces. Each topic has its own temporal quality directing the performers and audience to create associative meaning between events of the past and the present. For example, *keñes* pieces, as a type of *qoñır* piece, always function to command the audience's attention at the beginning of a performance. During the gathering, the audience can often tell "whose *qoñır*" it is, meaning which historical or folkloric figure, by listening to the opening part of the piece, and so can quickly focus their minds on the corresponding historical period. For example, Amire Ata in the story of "Ertis Tolnıqı" (The Waves of Ertis River) refers to the 8th century when the Scythians were active; Qorqyt Ata refers to the 10th century when the earliest Kazakh folklore and legends came into being, and Aqımcan refers to the 16th century and the birth of a grand Kazakh *ulus* (a pre-modern type of nation) across the steppes.

Along with the historical period, audiences may also be able to discern the topic of the piece from its opening: perhaps nature, animals, tragedy or lamentation. In this sense, *sybyzghy* pieces bring the various members of a gathering together giving historical personality and ethical or moralistic quality to the on-going performance. I would summarize the primary topics as follows:

1. *küi bası* (header), serving as an introduction to a series of pieces;
2. *tolğau* (meditation), usually describing an ancient figure's emotional state;
3. *zar* (tragedy), usually following individual accounts of personal hardship;
4. *sağınış* (absence), serving as a commemoration of elders, parents, or leaders, occasionally including pieces to commemorate leaders such as Mao Zedong and Zhou Enlai;[4]
5. *coqtau* (lament), describing someone's lamentation;
6. *oqyğa* (event), depicting historical events;
7. *keñes* (conversation), the image of preparing for dialogues on celebrities;
8. *mahabat* (love), telling love stories set in an indeterminate time and place;
9. *hayuanat* (animals), depicting animals in motion, sometimes in a series of companion pieces;
10. *tabyğat* (nature), *sybyzghy* performers' projections of rivers, mountains, clouds and so forth.

In gatherings, people play *sybyzghy* pieces according to the immediate topics that arise in the course of the conversation, rather than a selection based on musical content for entertainment purposes. The performances move and shift between subjects, of which the selection and progression are generally random and improvised. There are indexical terms that are rarely stated explicitly, but are rather implied through subtly coded social cues influenced by emotion and the affective properties of sound. Although the narrative transitions vary from case to case, these examples illustrate the strategies of linking the mood of the *sybyzghy* piece with the mental states implied by the narrative topics, signifying new or continued moods in the narratives of the gathering. These indexical terms typically represent the social and interpersonal state of mind of a gathering, while also comprising an aesthetic structure or order with which to direct and maintain the atmosphere.

In summary, a whole event gathering can be seen as a system of indexing, "linking" or "suggesting" meaning through sonic symbolism. The symbolic references are interconnected in certain orders, linked and hinted at by using indexical narratives that recall particular themes, while each theme has its own temporal quality leading to a unique atmosphere of imagined time, converting *a* past into *the* past, linking *the* past to issues of the present. For

4 Most performers regard these pieces as an integral part of the repertoire, rather than aberrations. They would play them in an informal gathering.

audiences, their conversational manner and the topics discussed are inflected by the established musical mood and empowered by a temporally constructed ethos of history. For the performers, the lyrical content and their choice of pieces to play are similarly decided by this indexical framework. In this way, it is the overall indexical and temporal framework of *sybyzghy* performance, rather than any isolated aspects of the performance itself, that constructs this unique type of meaningful communication.

3 Temporality in Musical Symbolism

As Becker and Becker (1981) remind us, musical systems or languages are "always more than organized sounds, vocabularies, and syntaxes" (215). Given this, and looking at musicality in human experience, what forms the quintessence of a musical performance? Through its groove[5] and atmosphere, a musical performance often leads us into a unique temporality in which our experience of musical time stands apart from our usual perception, prolonging or condensing our moment-to-moment experience, or sending our minds wandering in a world of personal reminiscence. Particularly in the case of *sybyzghy* performance, the multi-layered nature of time that is experienced serves as both entertainment and solemn ritual. In my article, "Painting with Sound: Performing Cultural Memory through Kazakh *Sybyzghy* Cassettes in Northern Xinjiang" (Wei 2024), I showed that the sonic metaphors and performative fidelity of recordings are meaningful in creating a present identity drawn from the historical past of Kazakhs. Throughout the research, I noticed that this relationship is built on the heavy fusion of music and narrative, which leads to, in Becker and Becker's words, "naturalness of music" along with proper "workings of time" (207). Over the hundred-year-long history of performing *sybyzghy*, performers and participants have collectively developed a unique genre of speech and musical performance to narrate their history, and more importantly, to create a trance-like atmosphere, immersing gatherings in recreations of historical time and invoking moral and ethical principles from these narratives to help construct the current identity of the Kazakh clans.

5 In ethnomusicology, groove defines the sense of rhythmic patterns, usually felt and described in generic terms of musical affectivity, which can be fundamentally defined in specific temporality of musical sections, events, and motifs. See Charles Keil and Steven Feld's (2005) approach to grooves on "motion and feeling," and Mark Abel's (2011) approach to grooves on temporality in his dissertation, *Groove: An Aesthetic of Measured Time*.

Historically, *sybyzghy* pieces cover a full range of topics, and *sybyzghy* performers use melodies, timbres, and speeches between pieces to represent emotive images of historical events, figures, and moral or ethical virtues; they then use these emotions to express ideas of identity and justice in the present day as well. In this sense, certain meaningful figures become symbols, e.g., names of notable rivers, heroes, and animals, which represent the Kazakh nation. Melodies and timbres present these symbols within the context of a particular atmosphere, or vision. The gathering as a whole responds to and is shaped by the way these icons and symbols of cultural memory are invoked. For most listeners, of course, such signs are typically not processed in terms of language-based thought but are simply *felt*, due to "a direct identity established by the resemblance between the musical signs and other expressions of excitement" (Turino 1999, 227). There is, accordingly, a general version of a *sybyzghy* piece, played with references to a generalized picture in the past, while real-life performances provide additional contexts that elaborate upon this general picture. In this sense, time itself—whether folkloric, historical, or mythological—can be invoked in a symbolic way by the performer, taking the audience into a direct experience of the values and meanings associated with each temporal context. This way, time itself, as invoked via performance, works to link or 'index' a relationship between past narratives, events, or notable figures and issues being discussed by the community in the present day.

The symbology of *sybyzghy* performance is interconnected through particular orders, and linked and hinted at through the use of an affective language. In Turino's (1999) terms, indexical relations are "grounded in personal experience; the members of social groups will share indices proportional to common experiences" (235). In the next two sections, I will give examples of temporally constructed visions of history, and show how these narratives then create a symbolic platform used by participants to map past times.

4 Historical Figures, Imaginaries, and Their Times

Like many other narrative musical forms, *sybyzghy* performance draws from historical tales, but here the figures and imaginaries are unique to Kazakh identity. In this section, where I illustrate the spectrum of temporality constructed through *sybyzghy* performance, I present the stories in a composite style, paraphrasing various written and oral sources. These sources are primarily retrieved from Ábdiqadir-úlı (1991) and Abdolla-úlı, Baymúrat-úlı, and Quanbek-qızı's (2009) writings, who compile their stories from various oral sources, as well as from my personal interviews with *sybyzghy* masters and event participants conducted between 2012 and 2017. I categorize three different groups according to their attributes of time: folkloric time, mythological time, and historical time. I also italicize the key temporal reference in each folktale, to show the general and specific tones that construct its particular temporality.

4.1 *Folkloric Time*

According to oral literature, the *sybyzghy* was created by shepherds while grazing, making a *múñlı* (sad, melancholy, or sorrowful) sound. Some folktales take the following form:

> During *a time of many hardships* for the Kazakhs of the steppes, a cruel *bay* (rich lord) sent two orphans to watch over a flock of sheep and protect them from wolves, threatening to beat them if they failed in their task. *One day*, the younger boy thought he heard the howl of a wolf and drove the sheep to a new location. *Over several days* he heard the howl again and again, until he realized that the sound was not a wolf after all, in fact it was the wind blowing through hollyhock reeds that had been eaten by the sheep. Plucking one from the ground, he began to blow into it, putting his fingers over the various holes in the stem to change its pitch. *One night*, the other boy heard this strange new sound and asked, "Are you crying?" The younger shepherd boy replied, "No, I am playing a melodious voice." *Since that day*, the boys began playing the hollyhock reeds together, creating new sounds. *Later*, the practice spread to the people, and the hollyhock came to be called '*sızılğı*'. *Over time*, folk scholars adapted the reeds with two, three and four holes to make new sounds, and this came to be called *sybyzghy*.
>
> Interview 2014

Such narratives are common to find if one travels to Kazakh communities in Xinjiang. The tone is vague in terms of time: phrases such as "a time," "one night," "over time," and "several days" create a sense of folklore, here appearing in a more general context in order to introduce an outsider, myself, to the culture.

On some occasions, *sybyzghy* pieces narrated in folkloric time are set within a more specific, but still vague, temporality, making reference to and mingling multiple conflicting historical times, according to the motivation of the performers. For example, the piece, "Aqsaq Qoñır" (A Crippled Brown Sheep), can be attributed to any time of hardship from the mid-18th century to the mid-20th century. It runs as follows:

> There was once a village on the other side of the Altay Mountains. *Threatened by dangerous times*, its people decided to relocate in order to protect themselves. Before the villagers could begin their migration, a sheep belonging to the head of the village was badly injured by a falling mountain rock, crippling one of its legs. This sheep was cherished by the village leader, who thought the injury must surely be a bad omen for the journey to come. So, he bandaged the sheep's wounds and left it behind in the village. Before departing, he considered slaughtering this brown sheep and taking the meat for the unforeseeable journey ahead. Yet he also knew that there was a superstition among the people that "slaughtering crippled sheep will ruin your possessions." After the villagers left, the crippled brown sheep followed the footprints of its herd who had left along with the villagers, and finally caught up with the migration team after a month. People were excited to see this sheep again. *Musicians were moved by the crippled sheep's persistence and loyalty to its master and the villagers*, and created a musical tune to celebrate its painful journey. This particular piece depicts the moaning of the crippled sheep as it follows the footprints.
>
> Interview 2016

Another piece, "Calğız Cigit" (Lonesome Man), offers the following:

> *Once upon a time in Altay*, a wealthy man died, leaving behind a son and a daughter who could play *sybyzghy*. The daughter thought that she might have to leave her brother one day after her father died, so she was very sad. Then she said to her brother: "one day I will leave you too, and you will be the only one left here. At that time, there will be wicked people robbing your property and even threatening your life. I suggest selling the livestock and turning it into gold and silver, so that we won't attract the attention of others." However, the elder brother did not listen to the advice of his wise younger sister. One day, he came to the pasture of the horses and found that the livestock had been stolen. He followed the footsteps of the horses to the territory of the thieves, took out the *sybyzghy* and played "Calğız Cigit" (The Lonesome Young Man). Later that night after noticing that her brother had failed to return, the sister walked to the pasture and found that the horses were gone. In her sadness, she played "Múñlı Qız"

> (Sorrowful Girl). The group of horses tried to escape from the enclosure after hearing the sound of the boy's *sybyzghy*, but could not find their way out. One of the black horses in the group was the boy's favourite horse. Upon hearing the boy's voice drifting over with the sound of the *sybyzghy*, he broke the rope and dragged the thieves with him straight through the fence. *From this incident, the song "Qara Attıñ Şabısı" (The Gallop of a Dark Horse) came into being*. Next, the jujube-coloured horse which the girl hand-fed also escaped and returned to her safely. "Tayküreñ" (Purple Red Horse) became a song about the horse returning to its mistress. The brother and sister rode their favourite horses and left their hometown, and spent the rest of their lives in a place called Búlğın, on the top of a yellow forest. There are now three hillsides, said to be the petrified siblings and two steeds. *These stories and music were played by Qoşanay when he was 85 years old*, who told the stories to Orazbek Qanapyn. Orazbek recorded it and told us.
>
> ABDOLLA-ÚLI, BAYMÚRAT-ÚLI, and QUANBEK-QIZI 2009, 278

In this piece, one can vaguely guess that the time of the story was after the mid-17th century, the time when Kazakhs migrated into the Altay Mountains. Performers, however, can locate the story in different times, according to their performance context, and can also add the present time into their narration, for example, in both pieces, lines like "musicians were moved by the crippled sheep's persistence and ..." and "these stories and music were played by Qoşanay when he was 85 years old ..." provide the listeners with a sense of the transformation of time. This leads as well to a connection between past and present which makes the performance seem more lively and natural.

4.2 *Mythological Time*

National communities have been known to adopt discourse around a particular 'ancient' reference point, such as a musical instrument or notable figure, to establish a history of their nation that can be traced to a time before the formal emergence of the national identity. In *sybyzghy* performance, this historical depth comes from constructing the temporality of mythologized historical figures such as Qorqyt Ata and Asan Qayğı. The complexity of 20th century stories about the early mythological history of *sybyzghy* reflects a diverse, cross-regional array of folklore throughout Xinjiang and Central Asia, and across a wide range of history. This construction of temporality contributes to the nation-building of Kazakhs in modern times as the setting and remote timeframe of the pieces helps to emphasize the long history of the Kazakhs, while they also reference the specific narratives definitive for the cultural life of Kazakhs today—most importantly in temporal references to historical figures and religious (Islamic and pre-Islamic) life.

Talğat Káripcan-úlı Múqışov, a professor at the Kazakh National Conservatory, reports another folktale version of the *sybyzghy*'s origin, but one that now references Alexander the Great and the Macedonian era:

> On the head of Alexander the Great, there was a pair of horns, and Alexander had been hiding this fact all his life. Whoever cut his hair would have to die to keep the secret. When Alexander was a child, a prophet once said to him: you will become a king in the future, and then there will be horns on your head. A pair of horns will grow, and when the horns are found, you will die. In Alexander's time, horns were believed to represent power and sovereignty. One day, a young boy cut Alexander the Great's hair, and, as usual, the boy was marked for death. When the boy was thrown into a lake, he found that the hay (dried, hard-stemmed grass) on the bank could make a sound. It turned out that insects had bored a small hole through each piece of hay, and it would make a sound when the wind blew. Noticing this, the boy grabbed a straw of hay and blew into it. A group of hunters who were hunting not far away heard the sound and found the boy. The boy was rescued, and he told the hunters the well-hidden secret of Alexander. As the prophet said, the great Alexander *then* passed away. The hay with a hole became the prototype of *sybyzghy*.
>
> Interview 2016

In contrast to folktales about pre-Islamic saints, such as Qorqyt and Asan Qayğı, mythological stories are set in a Biblical style of narrative; the general style is more a solemn recitation than casual improvisation, detailing the specific instrument that is mythologically symbolic of the nation, and featuring a time duration and procedural plot that indicates a holy or scriptural story. For example, in roughly the 8th to 10th century, Qorqyt was a famous poet, philosopher, musician, epic chanter, and high-ranking *baqsı* (shaman) in both the real Turkic world and folk literature. The historical Qorqyt was born in the Syr River region, and served as the prime minister of five khanates in his lifetime. According to oral tradition (interviews, 2012 through 2017), Qorqyt's mother experienced a three-year pregnancy. When he was born, after three days and three nights of 'confinement', his mother named him Qorqyt, meaning "to make people afraid" in the Kazakh language. Even as a child, this mythologized narrative goes, Qorqyt was very intelligent and could predict the future, thus he was regarded as a prophet by his tribe. While his deeds may be embellished or fantastical, the life of Qorqyt is placed within a specific time frame according to the composite, mythological stories of Qorqyt,[6] a compilation of oral tradition which holds that Qorqyt and the Prophet Muhammad lived in the same age. The Kazakhs believed their 'first and fundamental' instrument, the *qobyz*,[7] was invented by Qorqyt, who also left a large musical repertoire to the later Kazakh nation. Most literature on the musical legacy of Qorqyt begins with the prophet's fearful dream:

> *One day*, Qorqyt saw a group of people digging a tomb. He asked them, "Who was the tomb dug for?" The diggers replied, "This is Qorqyt's tomb." Qorqyt then woke from his dream with much trepidation.

The story continues:

> Qorqyt was searching for immortality and trying to escape from death, he visited all four parts of the world. But wherever he appeared, death would meet him in a different guise. Then he returned to his homeland, the shores of the Syr-Darya River, and invented the *qobyz* from a whole piece of wood, covering the instrument with the skin of his sacrificed camel, Jel-Maya. Thinking that death would not reach him on the waters of the Syr-Darya, he spread out a carpet on the river's waves and came there to play the *qobyz* day and night. His playing attracted all earthly creatures, animals and birds, who gathered enjoying the music, and so long as he played death could never approach him. *After forty days* ...
>
> JŪBANOV 2002, 280

Exhausted from his constant musical vigil, after forty days Qorqyt fell asleep, and Death finally turned into a water snake and bit him with its venomous fangs. Before dying, Qorqyt asked people to bury him by the Syr River and place

6 See manuscripts and publications under the name "Book of Dede Korkut" (or other similar titles) for renditions of the mythological stories of Qorqyt.

7 *Qobyz* (a.k.a. *kobyz* or *kylkobyz*) is a bowed instrument used by several Turkic groups that have a common nomadic history on the Central Steppe, including (primarily) Kazakhs, Kirghizs, Karakalpaks, Bashkirs, and Tatars. The instrument has two horsehair strings, producing, like the *sybyzghy*, a bi-phonic sound of two parts. Historically or mythologically, the *qobyz* was used by shamans (*baksı* in Kazakh) for healing and to communicate with holy spirits. Musical pieces about ancestors playing both *qobyz* and *sybyzghy* may cross-reference each other, pointing to a set of stories from the same historical periods with a common mythology, folkloric motifs, and morality.

his *qobyz* in front of the tomb. Nowadays, it is still a convention among the Kazakh nation to say: "Whenever you walk past Qorqyt's tomb, you can hear the voice of his *qobyz*" (personal interviews, 2012 through 2017).

In another example of constructing mythological time, Asan Qayğı Sábyt-úlı was a famous 14th-century Kazakh philosopher and *aqyn* (poet). He served as an important minister of the Golden Horde and the Kazakh Khanate. According to folk literature, he rode a white camel all his life, in search of a *cerúyiq* (paradise), which was rich in water and plants and free from feuds and oppression. In folk literature, Asan Qayğı worries about all matters concerning the life of the Kazakhs, from personal disputes to clan affairs. Growing weary of the cyclical nature of human joy and sorrow that he saw repeated all around him, Asan Qayğı rode forth on his camel to find a place where the cycle might be broken. *After ninety-nine days*, he came to the Mountains of Kökşe and found there a dense forest populated with many animals. Alongside them were men and women living, to his mind, an idyllic existence: a land without strife where the men toiled all day without rest while the women lived in comfort. Inspired by this vision of paradise he started singing: "*It rains during the day, it is hot at night*, and the land is restless; women are fat, men are thin, unfortunate men."

In the two examples, we can see how a narrative construction of temporality establishes a presence for Qorqyt and Asan Qayğı within the cultural imaginary of the nation that is fundamentally different from that of other historical figures. Their stories, often embedded in *sybyzghy* performance, are set in a Biblical style, invoking vaguely mythological times and asserting the longevity of the Kazakh nation. The usage of such mythological stories is still popular in Kazakh cultural life, influencing both expressions of wisdom and notions of history. As a result, the expressions are inherently connected with temporal terms; with people using alternative titles, such as "Qorqyt escaping life for forty days" or "Asan Qayğı's two years searching for paradise" in their folk narratives, demonstrating the role of these temporal terms in lending stories an air of gravitas and holiness.

4.3 *Historical Time*

Some *sybyzghy* pieces, often ones that portray the prowess of heroes and their horses, are narrated with specific reference to historical time. *Sybyzghy* tunes derived from relatively concrete historical accounts, reaching back to the era of Ghengis Khan in the 12th century, also highlight the shared cultural roots of merged tribes of Kazakhs and Mongols in the Altaic region. For example, "Aqsaq Qúlan" (Crippled Red Horse) presents a tragic, but philosophical, story of the Mongolian Khan and his son:

> Genghis Khan loved his son Jöchi, so much so that he ordained that anyone who brought him news of the boy's death would have their head filled with lead. One day, the poet Ketbuqa had a nightmare of Jöchi tumbling from his horse and dying, a vision that proved to be correct. To inform the Khan, Ketbuqa played his *dombyra*,[8] making a sound like a galloping horse. Upon hearing the song, Genghis Khan trembled and wept, asking, "Why does this song make me feel so sad, as if it heralds the news of Jöchi's death?" The poet put down his *dombyra* and explained the story of Jöchi's death in a poem. The Khan was ready to kill Ketquba in the manner he had ordained, until the poet countered: "It was not I, but the *dombyra* who told you of your son's death through sound." True to his word, the Khan filled the *dombyra* with lead. In modern-day Kazakh legend, this is how the *dombyra* got its soundhole.

Although the story could of course be apocryphal or invented, it does give details of real persons and 'real', that is to say not fantastical or supernatural, plot details. Unlike mythological or mythologized figures such as Alexander, Qorqyt, and Asan Qayğı discussed above, Ghenghis Khan, his son, and his talented minister appear in genealogy records and are considered actual ancestors of the Kazakh clans. Rather than suggesting vague or indeterminate times within Kazakh history, such figures and their stories invoke real historical power and identity that continues to affect social lives, including gathering events featuring *sybyzghy*.

Ghenghis Khan-like figures comprise a group of subjects narrated in *sybyzghy* performances, and as the stories become more realistic, more *sybyzghy* players appear in them as model figures for contemporary players. During the Ablay Khan era in the 18th century, the *sybyzghy* tradition became highly developed in the Syr River region. Berdiqoja was a famous *sybyzghy* player of this age, who served Ablay Khan by playing tunes about historical heroes and their martial exploits. In the 19th century, Cılqışı Ahtan-úlı became the first to have the double role of *sybyzghy*-master and local leader, and was highly respected for his ability to solve village problems through gatherings that featured the playing of the *sybyzghy*. Continuing into the late 20th century, the historical persons and events referenced become more and more 'real', and more connected to the socio-cultural and political milieu of the present, as they are increasingly used by the performers to associate their power and

8 The *dombyra* is a plucked string instrument common to Kazakhs.

social relations with current events. A trilogy piece, "Marğabıldıñ Qara Qasqa Atınıñ Şabısı" (The Running Posture of Marğabıl's Horse), portrays a historical horse-racing gathering among the Kerey and Naiman tribes of Xinjiang Kazakhs. It highlights the historical rivalry between the two tribes and the sense of prideful superiority the Kerey feel over the Naiman, a sentiment that still prevails among Kerey cultural insiders, even those who live hundreds or thousands of miles from their homelands. The story is as follows:

> Marğabıl [a Kerey] was a poor man who lived in the Altay area. He had a black horse with a white mane. One year, Imahbi of the Naiman tribe invited 12 Kerey tribes to come and participate in a festival of wrestling, horse racing and other activities held on the grand anniversary of a memorable event. The horse racing champion would receive nine young camels. Kerey's leaders hired trainers to select the horses for the race. The fastest horse in the Kerey tribe was one old grey horse, but the trainers were worried, saying "The old grey horse is old, not fiery like when he was young, and he may not be able to cross the finish line first."
>
> The Kerey leaders were concerned that people would be upset if they lost the race. At this time, a famous horse trainer stood out and asked to have Marğabıl's white-maned black horse and train it together with the old grey horse. When the day of the race came, the Kerey leaders took the two horses to the track. The Naiman leader, Imahbi, sent his horse trainer to observe the opponent's horse racing. After the man returned, he told Imahbi that the white-maned black horse was born in the place where the "banished female patron saint" played, and no horse could surpass it. Imahbi heard that and judged that the Kerey's horse was going to win. In order not to lose to the Kerey, a Naiman horse trainer planned a conspiracy in private. He set three traps on the track for the Kerey horse. In the end, the white-maned black horse and one of the Naiman's horses crossed the finish line tied for first place. The Kerey leaders then learned that the white-maned black horse had been sabotaged, and they clashed with the Naiman in anger. The impartial Imahbi did justice, rewarding the Kerey's two steeds after the truth of the matter was revealed, and giving them nine camels as an apology. People were very satisfied with Imahbi's approach, and the Kerey were full of admiration for the white-maned black horse, so they composed this piece of music.
>
> ABDOLLA-ÚLI, BAYMÚRAT-ÚLI, and QUANBEK-QIZI 2009, 68

Such pieces set in historical time are often played with narration tracing family and regional history, in an atmosphere that is specific to family, regional, and clan identity. When passing down narratives of *sybyzghy* pieces from generation to generation, people inherit more than merely the musical sounds and their link to historical memory; the authority of interpreting the meaning and significance of these events, each in its own temporal measure, is also passed down to younger generations.

5 Conclusion

In this article, I have explored the way that a collective understanding of temporality in *sybyzghy* performance rests on the fusion of narrative and sound. I provide examples of contemporary *sybyzghy* pieces that link to both historical and imagined times, and explain how, in terms of indexicality, this temporality is conveyed within a performance. I listed three types of narrative time—*folkloric*, *mythological*, and *historical*—performed in the narrations within contemporary *sybyzghy* performance. People employ these temporal frameworks to express a plurality of meanings within pieces, confirming their identity as descendants of ancestors who interacted with mother nature (as in the *Folkloric Time* section), as part of an ancient and storied nation (as in the *Mythological Time* section), and expressing pride in their clans in modern times (as in the *Historical Time* section). These diverse meanings illustrate the depth of cultural narratives by allowing performers to articulate these different visions of time as key aspects in a continuous process of nation-building and affirmation of present-day Kazakh identity. Through a process of indexicality, or symbolic reference, *sybyzghy* events engage unique historical times and figures, while each topic in turn conveys a temporal quality that forges links between the past and the present, pointing the way in the decision-making obligations of communal gatherings. This temporal quality, derived from the "lexicalized" concept of time in Abdramanova (2019)'s terms, defines the "moving ego" that encompasses a plurality of ideas from which Kazakh ancestral concepts are formed through the narratives of multiple times and their heroic figures.

In conclusion, I argue that narrative *time*(s) in Kazakh *Sybyzghy* performance, its historical figures, imaginaries, and its temporal framework form the quintessence of *sybyzghy* performance (and many other Kazakh expressive arts). Time forms a key aspect of Kazakh artistic and cultural life, shaping participants' sonic identities in the ways they experience time personally, interiorly, and emotionally (evoking Adams 2020). In the future, with my Kazakh colleagues, I will continue to explore what specific meanings, in Feld's (1984) terms, "from an inner form to an outer social realm," have been performed to construct the contemporary Kazakh identities that are already shown meaningful in the "workings of time" (Becker and Becker 1981).

References

Abel, Mark. 2011. *Groove: An Aesthetic of Measured Time*. Ph.D. Dissertation. University of Brighton.

Ábdiqadir-úlı, Múqamet. 1991. *Sybyzghy Sazı* (The music of sybyzghy). Ürümchi: Xinjiang qingshaonian chubanshe 新疆青少年出版社.

Abdolla-úlı, Qabılaldı, Quanış Baymúrat-úlı, and Gülnáz Quanbek-qızı. 2009. *Qazaq Auız Ádebyetiniñ Qazınası: Küy Añızdarı 1* (Kazakh folk literature anthology: Folktales of küi, vol. 1). Ürümchi: Xinjiang renmin chubanshe 新疆人民出版社.

Abdramanova, S. 2019. "Conceptualization of time in Kazakh Idioms." *Вестник КазНУ. Серия филологическая* 173, no. 1: 4–12.

Adams, Margarethe. 2020. *Steppe Dreams: Time, Mediation, and Postsocialist Celebrations in Kazakhstan*. University of Pittsburgh Press.

Becker, Judith, and Alton Becker. 1981. "A Musical Icon: Power and Meaning in Gamelan Music." In *The Sign in Music and Literature*, edited by Wendy Steiner, 203–15. Austin: University of Texas Press.

Beknazarova, Ulzhan. 2022. "Linguistic Metaphor of Time Concept in Kazakh Cognition." *Bulletin of the Karaganda University. Philology Series* 105, no. 1: 102–9.

Feld, Steven. 1984. "Communication, Music, and Speech about Music." *Yearbook for Traditional Music* 16: 1–18.

Daukeyeva, Saida. 2015. "'Aqqu' (White Swan): Sound Mimesis and Spirit Evocation in Kazakh Qobyz Performance." In *Pieces of the Musical World: Sounds and Cultures*, edited by Rachel Harris and Rowan Pease, 67–86. New York: Routledge.

Keil, Charles, and Steven Feld. 2005. *Music Grooves: Essays and Dialogues*. Tucson, Ariz.: Fenestra.

Lias, Godfrey. 1956. *Kazak Exodus*. London: Evans Brothers.

Janabekova, A. 2012. *Qazaq Tilindegi Funktsionaldi-Semantikalïq Temporaldïq Qategoriyasï* (The Functional-Semantic Temporal Category in the Kazakh Language). Almaty: Eltanïm.

Jubanov, Akhmet. 2002. *Ghasyrlar pernesi: Qazaqtyŋ Khalyq Sazgerleriniŋ Ömiri men Shygharmashylyghy Turaly Ocherkter* (Strings of the Centuries: The Lives and Works of Kazakh Folk Composers). Almaty: Daik-Press.

Pegg, Carole. 1991. "The Revival of Ethnic and Cultural Identity in West Mongolia: The Altai Uriangkhai Tsuur, Tuvan Shuur and Kazakh Sybyzgy." *Journal of the Anglo-Mongolian Society* 12 (1–2): 71.

Pegg, Carole. 2001. *Mongolian Music, Dance, and Oral Narrative: Performing Diverse Identities*. Seattle: University of Washington Press.

Pegg, Carole. 2021. "Cradle of Drone-Overtone and Timbre-Centered Music: Cultural Landscapes of the Indigenous Peoples of the Altai Mountain Range and Its Neighboring Areas." *Asian Music* 52 (2): 11–45.

Pegg, Carole. 2024. *Drones, Tones, and Timbres: Sounding Place among Nomads of the Inner Asian Mountain-Steppes*. Urbana: University of Illinois Press.

Peirce, Charles Sanders. 1955. *Philosophical Writings of Peirce*. Edited by Justus Buchler. New York: Dover.

Post, Jennifer. 2024. "Making and Growing End-Blown Flutes in the Mongolian Steppes." In *Instrumental Lives: Musical Instruments, Material Culture, and Social Networks in East and Southeast Asia*, edited by Helen Rees, 137–76. Urbana: University of Illinois Press.

Samuratova, Tatigul Kakenovna, Temirzhan Kumkhametovich Musalimov, Nazymgul Albytova, Mukhtar Makhanov, and Nasgul Kakenovna Omarbekova. 2015. "Space and time as the fundamental elements of Kazakh World Image." *Mediterranean Journal of Social Sciences* 6, no. 5: 127–32.

Wei, Xiaoshi. 2024. "Painting with Sound: Performing Cultural Memory through Kazakh *Sybyzghy* Cassettes in Northern Xinjiang." *Asian Music* 55 (1): 4–37.

PART 3

Time's Uncertain Measure

CHAPTER 11

Temporal Saturation; Whens

Chloe Garcia Roberts

Abstract

The following prose poems recount instances of a phenomenon I have termed "temporal saturation," a proposed measurement of how certain moments in a life when remembered seem to spill or shrink, to transcend or subvert their physical duration, and color differently their surrounding time. It is a measurement not of physical increments but of emotional density, qualified by the vividity by which these moments are reexperienced, relived, and returned to the present. I am interested in the resonance of these moments of temporal saturation, not only in how they reverberate through an individual's life, but also in how they may resound with undiluted intensity through the lives of others.

1 Temporal Saturation

Memory's inscription on the mind can be compared to afterburn on a retina after a flash of light. It is a physical enlightenment measured by the degree to which the gesture, the occurrence, the after-living, haunts the individual in blue repetition until they are released. The depth of this inscription and variations in its amplitude are a result of the phenomenon of temporal saturation—the explanation for why certain moments of your life seem to spill or shrink, to transcend or subvert their physical duration, and color differently their surrounding time.

Temporal saturation is an elusive measurement disproving any correlation between quantity and influence, that is used to explain both the canyons that can appear inside moments of great rending, joyous or horrific, entombing an incarnation of the self which will never again exist; as well as the median intervals of floating passivity that resist recollection and whose ending is marked by a feeling of awakening: a drowsy startle or a gradual reconsciousness.

High levels of temporal saturation are evidenced by a languorous stretching of the experienced present, which then refracts and amplifies the emotion of the moment. The joy making this spreading pleasurable, the fear terrible,

 | DOI:10.1163/9789004723634_013

though both poles can be described in terms of the sensation of falling. The difference being that the first is a falling *into* and the latter a falling *through*.

Low levels of temporal saturation are evidenced by malaise, an involuntary refusal on the part of the individual to knit themselves to the place they occupy. Homesickness—the corporeal and spiritual longing for a physical and temporal point of greatest belonging—is the best diagnosis to describe these ebbs of existence.

•••

So, the instance when in your room one afternoon you first reached for my face, placed your hands on either side and sighed. A sigh not of air escaping the lungs, but space escaping the heart—which suddenly I filled. And in that displaced space, that love broke around me, I drowned, I drown, I will forever drown. I burned, I burn, I will forever burn.

And conversely the feverish restless years I spent in the city, insensate, starving, trying always to convince myself that I loved things I did not. The night I called a boy, a beauty, talking to him through the wide night window of the city, talking to him and to the city itself, arguing about who would come to who across those streets, who would burrow crosstown through the inertia. Finally deciding to do nothing. We hung up. I left shortly afterward, and years later saw him at a party where he told me how the part that called me and that I weakly tried to answer was dead, that he had killed it. This isn't an anecdote of the one who got away. It is evidence of a state and the proof: that despite our magnetism we both succumbed. And while some might posit that that was just the result of our indifference towards each other, the truth is we were both living through, which is a definition of surviving. No thing presented us with the hard edge of reality, no one acted on us with opposing and equal force, certainly not each other. (Is that youth? I no longer know.)

The measurement of temporal saturation then can be used to quantify both the abscesses and the vividities, these gestures floating in great swathes of meaningless automation. Just as the atoms composing a human body can condense smaller than the head of a pin, the self can like a black hole collapse, like a poem reduce. And the proof of the emptiness which oceans those bright livings is how they sparkle and call to each other despite the expanse of the interims, be they seconds, decades, lifetimes. Inlaid in space, they form the constellation of the soul.

A final note: there will be certain people in the life of each of us, figures ringed in mothglow, who can be expressed as refrains of richest temporal saturation. They run like a weft through, becoming so inseparable *you* ultimately cannot be defined without *them*. If you can recognize these people in their approach, these are the figures you will navigate by, by whom you will plot your journey. They represent the closest approximation to knowing another being. Which of course, has been proven over and over again, to be impossible.

2 Whens

Remember when one morning we were home alone and instead of letting me brush out the tangles in your hair you asked to brush mine? Because you were so small, I sat cross-legged on the floor. At first the weight was too unwieldy for your tiny fat hand, and the brushing was more like being beaten softly about the head. But after a few minutes you got the hang of it and found your rhythm, rowing the brush roughly down the length of my hair. Lifting your arm to the crown of my head, letting it drift down again, lifting your arm to the crown of my head, letting it drift down again. Then, behind the slow shushing of the brush, behind the curtain of hair hanging around my face, behind my eyes closed underneath, an other side opened: a hallway into a when of you brushing my hair when I was no longer able. I felt just under the skin of my body, strong and capable as a mother, another body, weak and older. A self faintly felt of what was to come not what was. I felt the weight of both of your hands, the baby, the woman, my responsibility, my caretaker all in unison for a breath (and I do not say moment, or second because that implies a unit of linear time and this was not of that rubric but rather a transversal touching together of different times). I heard the sound of the brush, the sound of your breathing, the sound of the morning behind my eyelids, but in two places, two points on a life. Together, you and I, we had brushed open a whens.

The reason a certain moment can seem suspended from the stream has nothing in fact to do with the present in which it occurs, but is instead a result of the number of future instances that that exact moment is recalled, remembered, replayed. This phenomenon is actually just our future selves over-thinking the happening in question, which in turn breaks down its structure, its constitution like worn cloth or hollowed stone. This is the reason why the noise of life can sometimes silence itself, thin itself, to one tenuous thought like a silvery tendon spanning two dense slabs of muscled darkness (also known as the future and the past). Some people refer to this clarity, this one-note knowing, as intuition which is false, primarily because it does not come from within (in-tuition) but from without. It is not a product of what you are (i.e. what you have become) but of what you will become washing backwards. Whens then is a term for the tunnel between, the backward runnel which allows simultaneity to draw together, to bridge, two distinct times. So in this case I will be in a bed, in morning light, silent under your brushing, thinking, thinking, thinking backwards to that first time under your hand.

Eventually every life will prove the existence of innumerable tiny currents running like this one from the future to the past, even if they are largely undetectable and invisible, even if like lines of faint embossing they are easier felt than seen. And experience of this phenomenon, a face you remember loving before you saw it for example, is a clue one could use to determine that in this way the true future lies. To experience a whens is proof we live in a river that runs all ways, always.

•••

I've always thought that the best birth story I've ever read is that part in *Huckleberry Finn* when departing his own bloody murder scene, he unfolds himself into a boat and drifts into a mist shot with voices from the shore, this one and the other. Because he was in two places at once and also neither one; because he was between the shores; because he was a bridge or maybe better, he was becoming a bridge; because he was a whens.

At least this is how it was for me when I reached for you, when you came to me. Blind as a boat, floating on the currents of other voices, the waves of other's touch, the deeps of pain. When we were both listening for the other. When I was a body of water in which you were surfacing. When I was the sway, the stretch, the snap of molecules. And when from beneath you rose: Una ola, a wave. Una ala, a wing.

My own birth story, or the one that is not my mother's to tell, happened one afternoon several years ago in a blackout over the whole lower half of the city. I wanted to be home and foolishly decided to try and navigate the lightless hallways of my apartment building alone to get there. After climbing the stairs in muffled light from high and far off windows, I opened the heavy fire door to my floor, taking a quick look in the grey of what direction I needed to go. But when the door shut behind me, it shut so darkly it was as if it never existed. And when I stepped with fingers outstretched into the deepness, I was unmoored, eyeless. At first I refused to believe in my powerlessness and floundered forward. But with each step I lost myself a little more, feeling that strange vertigo particular to pitch blackness when the borders of the body start to inflate then reduce then waver. When finally I realized I was unable to reach my home, and that I was also incapable of returning the way I had come, I began to cry for help. And after a few minutes someone silently opened their door behind me and an oar of sunlight cracked the dark. I found myself just a few steps from my own door, in a corner. I had been asking for help from a wall.

For the rest of my life, as long as I am, I am there. Pressing myself against the faceless unanswering wall of god when behind me a door opens, a slight wing of light. This was a whens also, though a different type. Not a bridge connecting two points but the surge under the span—the flood.

•••

A decade ago I was driving with my mother across the country. As we drew closer to the Mississippi we began to hear on the radio that the river was swelling its banks and slowly spreading and cutting off all the passages back east. As we drove, we flew by empty houses next to the highway filled with dark water. Great mouthfuls were torn out of the countryside, and replacing that missing land was a waving stillness whose rising was so incremental, so unstoppable, it was answered internally with an equally black spill of fear.

When we finally made it to the banks of the great river itself, wide as a small sea, we drove straight across its turgid waters on a vast suspended bridge. And just under the surface of that suspended moment of crossing, I carried another moment: when less than a day before in northern Minnesota, I myself had spanned the infant river at its clear source in Lake Itasca, so slight and slender I had put one foot on this shore and one foot on the other.

A flood is just a widening of the between that causes a spilling of what was carried into what had carried it; an eclipse of the holding by what was held. *Après moi le deluge*, the French king says. And what he means is that he is the first singing sign of a reversal of the order of the world (another definition of to flood). But of course, we all can say this. The key is the word après, which yes means after, but also means below and most importantly, beyond—the direction whose mercy we all float along.

This is the moment where I admit to you that I have discovered time travel to be possible. And that I was ultimately taught this by translating the pain of a 9th century poet into English. To clarify, I do not mean that a body can move to a different time than the one they were born into, but rather that like a stitch one can gather not just all the shores the self has touched, but others that have been shown to you clearly enough that you may claim them rightly as your own.

Let me explain: we know the way fiction works is that the words cause certain chemicals in your brain to go through a mirroring dance without action that makes the rest of the brain feel as if what is read is happening to it. The way poetry works is similar, but instead of action it involves emotion. And this is where it gets a bit complicated, because emotion isn't mirrored or aped. Like light, it is perceived or it isn't. And if it is perceived or felt then it is original, inseparable from any other time that it is. So a man burning in grief can send an arrow of words forward in time, can through centuries, touch, can prove that years are not opaque but translucent, and thus can let those words, like light, pass through. Perhaps this is why to write is to dissipate the gloom, dilute the viscosity of the dark, because sometime, somewhere, now, someone else is lightening you.

•••

In the place of my childhood floods came like lightning, they were immediate, a flash, an issue, an opening of water in the desert, an apparition made real. The rivers there are born torrentially like gods, whole and fully formed. Once in the spring after the snowmelt, a friend and I went out to the arroyo to throw sticks into the whipfast current that had appeared overnight. Suddenly, I no longer know why, I stripped down to my underwear and stepped into the rushing water. I remember being hypnotized by the speed as I walked into its center, the way its force hit me, enclosing me in its power. And then where I had been standing, smiling, feeling the insistent pushing of the water against my legs, I was under, I was spun, I was thrown, down, around, away from the air. Time moves quick as water when you are about to drown, your flailing, your ragged breaths just so many droplets forced to rush in tandem along. Greater force holds you now, and not playfully in the least, tightens its hold.

Then just as I went completely under, I was thrown against a bank, a thin outstretched arm of sand, and suddenly I was free. I walked back upstream to my friend crying on the shore and we went home. My mother made a fire, wrapped me in a blanket, and while I sat cross-legged staring silently through the flames back into the mouth of the water, she brushed my hair.

CHAPTER 12

Once Upon a Time Again and Again: Temporal Non-Linearity in Graham Swift's *Waterland*

Sanyogita Singh

Abstract

This paper focuses on Graham Swift's *Waterland* and how its peculiar narratives suspend cause and effect. Rejecting a standard linear approach to time, the text instead manifests time through repetition and difference, as marked by the passage of water, in particular the river Ouse, as an indication of the constant and circular flow of time. The essay builds on Deleuze's ideas of difference and repetition to point to temporal existence in *Waterland* as not merely an infinite repetition, but also an eternal renewal. In this model of "difference and repetition" time appears not as an external force acting upon individuals, but rather as an internal and subjective construction, shaped by the interplay of repetition and difference that is inherent in memory.

Keywords

circularity – difference – literature – memory – nonlinearity – repetition – time

Temporality is the order which defines the experience of our lives. We exist within and have a relationship with time, more intimate than that which we have with space or mass. And yet time, as Carlo Rovelli writes, is "perhaps the greatest remaining mystery." Its evasive nature, seems at first "something simple and fundamental that flows uniformly, independently from everything else, from the past to the future," but later emerges as a superposition of multiple temporalities.[1] Time in literature also appears in various models and discourses: linear/non-linear, personal/public, individual/shared, subjective/objective, standardised/non-standardised. These multiple temporalities, often in binaries, are afforded to us by the "dialogic imagination" of a novel and

1 Carlo Rovelli, *The Order of Time* (New York: Riverhead Books, 2018), 11.

 | DOI:10.1163/9789004723634_014

the temporality inherent in its narrative.[2] What emerges is an aporia between perceiving time and measuring it. Does the experience of the passage of time change depending upon the observer's position? If yes, how does one articulate this difference, of false recognition, between different orders of time? It becomes evident, early on, that modernity's universal, linear time model falls short when it comes to addressing the multiple and mutable forms of time. Firstly, as Braudel points out, the world is not driven by a singular time. Elements have their own rhythms which may overlap but are rarely synchronised. Further, regarding linearity, one must reconfigure universal categories of "the past," "the present" and "the future." How does the present relate to past and future, and how do they differ from each other? Is present the gap between the past and the future? And more importantly, how do past and future relate to each other in a given present?

1 "WhyWhyWhy" *Waterland*?

Waterland (1983), intent on preserving a present tense of time, has time as its object of reference. The novel delves into local, personal, natural and global histories to discuss the fleeting here and now and through Fenland, its geographical setting, explores both fairy-tales and reality. For Tom Crick, the narrator, history is on the one hand opposed to superstition, and on the other appears as an amalgamation of realism and fabrication. Tom's narration of his various histories is often interrupted by the *now*, the moment when the reality of events sets in, where "Nothing" happens ("neither progress, nor decay," "no memory no history").[3] Swift thus depicts reality as "an empty but fillable vessel" one that may be filled either with the Here and Now or with history.[4] Swift uses the "Here and Now" to evoke the past and push the narrative forward; once his subjects have been engulfed in history, it suspends the regular passage of time. As he writes, "Here and Now had more than one face."[5] *Waterland* thus proceeds through the tense dialectical relationships between history and stories, history and histrionics, history and speculations, history and curiosity, history and memory, history and the Here and Now, continually rehearsing

2 M. M. Bakhtin, *The Dialogic Imagination: Four Essays*, trans. Caryl Emerson and Michael Holquist (1982; Austin: University of Texas Press, 2010).
3 Graham Swift, *Waterland* (London: Simon & Schuster, 1983), 6.
4 Swift, 42.
5 Swift, 58.

and modifying the definition of history as the narrative jumps back and forth in time.

History appears first in the novel in its epigraph, "Historia, -ae, f. 1. inquiry, investigation, learning. 2. a) a narrative of past events, history. b) any kind of narrative: account, tale, story." We then encounter the end of history, in the second chapter, in Tom's classrooms, where he is trying to "put himself into history," abandoning reality and searching for meaning via stories—"history as a fairy-tale."[6] What follows is a negotiation between history as an explanation of "WhyWhyWhy" and history's dissolution into yarns.[7] Additionally, Tom problematizes the idea of progress by intertwining it with the history of the world:

> There's this thing called progress. But it doesn't progress, it doesn't go anywhere. Because as progress progresses the world can slip away. It's progress if you can stop the world slipping away. My humble model for progress is the reclamation of land. Which is repeatedly, never-endingly retrieving what is lost. A dogged, vigilant business. A dull yet valuable business. A hard, inglorious business. But you shouldn't go mistaking the reclamation of land for the building of empires.[8]

Here Tom refers to the liquid landscape of the Fens, made from silt, always in the process of being reclaimed. Fens represent the coalescent texture of fairy-tales and reality embedded in history. The land reclamation process that is always afoot thus appears to unravel the mystery of the past, progress, and the constant merging of memory and time. In this way the non-linear narrative and story of *Waterland* pushes for a restructuring of the ideas of history, future, and progress. This inescapable repetition and continuous creation of difference leads to a transformation in our understanding of past, present and future.

Waterland narrates multiple histories—a family history, a history of the land, a history of human progress, a history of *Anguilla anguilla* myths—exercising the voice of a history teacher, Tom Crick. Similarly, memory is enmeshed in invariably connected narratives of histories, as history and memory represent different modes of dealing with the past and the Here and Now. Augustine sees memory as an aspect of time, ever-present in our minds. Carlo Rovelli considers memory a "solder" for processes that contribute to our making of reality.[9]

6 Swift, 12.
7 Swift, 100.
8 Swift, 302.
9 Rovelli, 100.

Chris Marker likens memory to history, an impossibility and a superimposition of images.[10] Jameson regards it as "temporal imprisonment in the present."[11] And, for Proust, it is memory that makes up reality. The discourse of memory in *Waterland* develops through a repetition of keywords such as remembrance, amnesia, and forgetting. Characters often labour against reality by dismissing the past, escaping causality by creating zones of interiority disconnected from it. Even in this quest to erase traces of the past, however, and to inhabit only the Here and Now, the narrative invariably links the present to the past, and perhaps even to the future. The effacement of memory thus works in dialectical opposition to the recollection and recording of history. As Swift writes, "since you cannot dispose of the past, since things must be—they had to make do."[12] In this way the attempt to forget a despairing past wrestles against a duty to remember, to carry on towards the future, turning memory into a metahistorical category. In other words, memory transports reality to the past while history develops in time for the present. Both imply a consumption of time and impinge on time's existential structure. According to Tom, "There are no compasses for journeying in time."[13] He describes every movement as going "in two directions at once. It goes backwards as it goes forwards. It loops."[14] The experience of multiple temporalities thus acts as a "double subterfuge,"[15] a dynamic temporal logic embodied in the non-linear narrative and brought to the fore by repeated metaphors and allusions to the River Ouse and political revolution.

2 Methodological Perspectives

For Derrida and (especially) Deleuze, repetition, as it is prior to the formation of identity, forms a focal point within poststructuralist inquiry. For Derrida, the repetition of the same in diverse contexts produces difference, thereby deferring the construction of a singular identity or meaning.[16] Deleuze discusses the concept across his oeuvre, starting with his early writings on Proust and

---

10 Chris Marker, *Sans Soleil* (Argos Films, 1983).

11 Fredric Jameson, "The End of Temporality," *Critical Inquiry* 29, no. 4 (2003): 695–718.

12 Swift, 110.

13 Swift, 122.

14 Swift, 122.

15 Russel West-Pavlov, *Temporalities*, The New Critical Idiom (Oxford: Taylor & Francis, 2012), 23.

16 Udaya Kumar, *The Joycean Labyrinth: Repetition, Time, and Tradition in Ulysses* (Oxford: Clarendon Press, 1991), 7. He writes, "In the texts of Derrida, repetition is seen as preceding the establishment of identities. Iterability becomes a condition of writing, an idea

Nietzsche, and later developed in his seminal work, *Difference and Repetition*. Like Derrida, Deleuze sees repetition as a linchpin in understanding how identity evolves. Moreover, he draws a distinction between two different modes of articulating repetition: naked and clothed repetition, with the former suggesting recurrence of the same as same, while the latter denotes change, i.e., difference. While naked repetition achieves the replication of an original identity, so that each occurrence of the repeated element is merely a duplicate of the original, clothed repetition makes such an original model hard to imagine. Plato's idea of forms as a model for objects suggests the paradigm implied by naked repetition, while the relation of original to copy implied here is erased in the idea of clothed repetition. Here each act of repetition affirms singularity—repetition is the technique of articulating difference without the "mediation of identity or analogy."[17] Further, Deleuze posits the special effects of difference through the concept of the différenciant, or the differentiation of difference. He writes, "[c]onsider two propositions: only similar things differ; only differences resemble each other. The first formula poses resemblance as the condition of difference; undoubtedly, it also requires the possibility of an identical concept for two things on the condition of their resemblance and also implies an analogy in the relation of each thing to its concept. Finally, it entails the reduction of difference to an opposition determined by these three moments."[18] What particularly interests us here are the prerequisites of an organisational structure that permits difference to be defined in this way. According to Deleuze, the first aspect of such a system is serial organization.

3 Jumping in and out of the Narrative, Non-Linearly

Deleuze distinguishes between repetition and resemblance from the outset. Whereas repetition is connected with distinction, resemblance is related to similarity. Hence, when a subject is irreplaceable or not interchangeable, repetition will occur. For Deleuze, this repetition is an infraction which signifies future-oriented thinking. In *Waterland*, in contrast, it takes numerous forms, including linguistic schemata, analogous circumstances, and, most essentially, thematic ordering. Recurrence adds flexibility to the narrative's otherwise non-linear arrangement and highlights the novel's narrative logic.

---

implicit in the notion of différance: iterability involves the repetition of the same as the different, inducing differentiation."

17 Kumar, 8.

18 Gilles Deleuze, *Difference and Repetition*, trans. Paul Patton (1994; repr., New York: Columbia University, 2004), 153.

Swift thus emphasises the associations between disjointed elements, arranging the narrative as an operatic event through repetition and rendering the plurality of time and its contours commensurable. I focus on two types of temporality here—history and memory—both of which underline Swift's non-linear temporal processes through repetition. This is not the "naked" repetition of character names, or places, things, and events which recur with the same meaning. Rather it is a repetition like the one in Tom's chronicle of the Atkinsons: "Which way do we go? Forwards to go backwards? Backwards to go forwards?"[19] Here, the use of 'forwards to go backwards' recalls an earlier use of the phrase: "Nothing moves far in this world. And whatever moves forward will also move back. A law of the natural world; and a law, too, of the human heart,"[20] just as the memory of the previous passage informs the present one.

Rather than stabilising or confirming the first occurrence of the phrase, the second appearance draws attention to the distinction between the two, a difference that occurs in both context and connotations. This type of repetition, which destabilizes identities and produces differences, appears frequently in Tom Crick's narration, as specific words and phrases from earlier are brought back, occasionally changed or perverted. In certain circumstances as well, the outward meaning of the expression is less relevant than the context in which it was first used as, in the repeated usage of "nothing." One such instance occurs as Tom narrates his father's memories:

> For a long time, he finds it hard to separate in his mind the familiar but foreign fields of the Fens and the foreign-but-familiar mudscapes he has come from. He expects the ground to quake and heave under his feet and become a morass. He is sent to a home for chronic neurasthenics. He thinks: there is only reality, there are no stories left. About his war experiences, he says: 'I remember nothing.'[21]

This particular passage and its "nothing" can be understood adequately only in reference to the context in which the idea originally appeared: "When you work with water, you have to know and respect it. When you labour to subdue it, you have to understand that one day it may rise up and turn all your labours to nothing. For what is water, which seeks to make all things level, which has no taste or colour of its own, but a liquid form of Nothing?"[22] Water and

19 Swift, 87.
20 Swift, 69.
21 Swift, 23.
22 Swift, 18.

landscape, which lie at the basis of the Fens, seem to be dismantling its residents. Tom's "nothing" thus echoes on multiple levels, especially with regard to human progress and productivity.[23]

Not all the instances of repetition in *Waterland* can be understood through the psychological recollection of its characters. In the recurrence of the phrase, "only important thing" for example, repetition serves as a tactic extending beyond character-subjectivity and memory:

> 'What matters', he went on, not knowing what sort of fairy-tale was about to envelop both his history teacher and his history teacher's wife, 'is the here and now. Not the past. The here and now—and the future.' [...] 'The only important thing ...'
>
> 'Yes, Price—the only important thing—?'
>
> 'The only important thing about history, I think, sir, is that it's got to the point where it's probably about to end.'[24]

4 Differencing with(in) Narrative

Waterland also contains instances of repetition even further from being subjugated to characters and their memories, extending to the methods by which Swift examines remembrance purely in terms of language.[25] The overture of Swift's chapter "About the Eel," introduces a succession of disconnected pieces that are then repeated throughout the text as part of understandable phrases. Some of the segments are already repetitions from previous episodes. "Once upon a time ...," for example, appears several times in "End of History," "Before the Headmaster," "An Empty Vessel," and "About the Change of Life." Here and in "An Empty Vessel," which contains a methodical repeat of prior events, now in the form of imagination, repetition occurs as an element of narrative organisation rather than as an element of a character's subjectivity. Thematic repetition forms yet another important element in *Waterland*. A multitude of symbols and images recurring throughout the text introduce and sustain a cyclical temporality.[26] The first metaphor for this circularity is

---

23 Tim Ingold, "The Temporality of the Landscape," *World Archaeology* 25, no. 2 (1993): 152–74.

24 Swift, 12.

25 Jennifer J. Smith, "Writing Time in Metaphors," in *The American Short Story Cycle* (Edinburgh: Edinburgh University Press, 2018), 70.

26 Mark Currie, *About Time*, The Frontiers of Theory Series (Edinburgh: Edinburgh University Press, 2010), 149.

revolution, which Crick notes is etymologically a "turning round, a completing of a cycle."[27] Swift's protagonist acknowledges that a revolution is "the last setting in which one would expect Nostalgia to thrive,"[28] yet Swift asserts that through this insurrection society rejects a continuation of history as a fabrication of the untrue and "decadent."[29] If the goal of revolution is to return to the past, then this sort of public uprising is not only a literal declaration of the tautological nature of time, but also a rejection of the Victorian paradigm of positive historical development. People desire the establishment of the past in the present during a revolution, and they want this history to serve as a guide for the future.[30]

Another negation of linear time in *Waterland* is seen in the use of the river Ouse, a body of water that rushes ahead through the Fens, history, and Tom Crick's tale while also twisting back on itself and repeating its journey. This stretch of water marks the passage of time throughout the story. Instead of just indicating the steady onward flow of history, the Ouse interjects itself into it, as water becomes a key player in the narrator's life. Water is the scene of Freddie Parr's murder and the ultimate resting place for Tom Crick's unborn child, taken from his soon-to-be wife's womb. On the other hand, however, water also offers a source of sustenance for the Crick family, who consider themselves as "water people."[31] The Ouse, according to Swift, "flows on, unconcerned with ambition, whether local or national. It flows now in more than one channel, its water diverging, its strength divided, silt-prone, flood-prone. Yet it flows. Oozes on, as every river must, to the sea … So that while the Ouse flows to the sea, it flows, in reality, like all rivers, only back on itself, to its own source. …"[32] Swift thus links the cyclical character of the Ouse with the circularity of history, but he also employs the river to call into doubt a view of history as proceeding through definite meanings and for fixed purposes. He claims that the Ouse runs without "ambition," bestowing both life and death on the inhabitants of the Fens and on Tom Crick's family throughout their ancestral heritage. In this way, the river circles back towards itself, but its impact on history and time is unclear. The Ouse calls into question the cause and effect model of historical evolution simply because it causes and effects a wide range of outcomes all at once. How this single body of water can simultaneously generate and demolish on its journey right back around is mysterious, surprising, and unpredictable.

---

27 Swift, 293.
28 Swift, 124.
29 Swift, 137.
30 West-Pavlov, *Temporalities*, 101.
31 Swift, 17.
32 Swift, 145.

5 Totality and Abstraction: Repetition, Memory and Narrative

In "About the Fens" Swift conceptualises the archaeology of *Waterland* from this point of view. It is a gestation and birth experience, enacted on a number of homologous levels: the embryo, the planet, language, as well as the prehuman, the human, and the posthuman. In most of these circumstances, gestation is not regarded as evolution in the sense of a seed gradually growing into its objective, a plant. It is regarded rather as differentiation, a sequence of differences or sets of differences that supersede one another. Swift's embryonic chart is thus nothing more than a collection of differences—a dark forerunner. This holds true for the other levels as well, especially in the case of language, where one set of stylistic qualities is replaced by another. The voice that imitates different styles creates not an identity, but rather a comparison of differences. The text of the episode fills in the gaps between these different series, establishing an endless process of interpretation that refuses to subsume the text under a single totalizing meaning.[33] The chapter's opening meditation reveals only stages of differentiation. Gestation appears as a repetition of life, beginning and ending in a state of chaos. This also demonstrates a framework of recollection, a remembrance that additionally implies an abandoning of identity and style. Like *A la recherche du temps perdu, Waterland refuses to reduce* previous experience to what it is purported to imply in the present. Both works recognise the structural interaction of past and present, bringing together a genesis realised by the protagonist and a structure discovered at the conclusion. Both books rely on looping designs in which the protagonist goes on to compose the first pages of the novel.[34] Even so, neither Proust nor Swift limit genesis to the structure it creates—the lived past will not be absorbed by the knowing present. Proust thus relies on memory and art, resolving the instantaneous intensity of the sensation with its later, more insulated, appropriation as knowledge. But while memory and art, in different ways, reveal us as essences that maintain, without merging, intensity and distance, the narrator's voice nonetheless retains its uniqueness through the numerous separate threads of time at work in its composition.

Swift uniquely tailors his book with respect to these temporal needs by rupturing the narrator's voice into multiple discourse zones, where intensity is experienced and meaning examined. Thus, the novel, a tale of articulations rather than one single articulation, employs the metaphor of water

33 Keith Ansell-Pearson, *Germinal Life: The Difference and Repetition of Deleuze* (Oxford: Taylor & Francis, 1999), 64.

34 Mark Currie, "The Novel and the Moving Now," *NOVEL: A Forum on Fiction* 42, no. 2 (2009): 318–25.

towards the end of each of its many parts. Articulation is seen as participation in a discourse, in a universe of meaning that will eventually be displaced by another reality. Memory, then, has a far different role in *Waterland* than it does in Proust. Memory, for Proust, combines intensity and the disclosure of an essence. It lends the language of the present resonance, allowing it to fluctuate between an anterior, which was once the lived present, and a future which is the present of Proust's writing.[35] Memory in *Waterland, in contrast,* conjures and reveals discursive remnants of a disarticulated reality. There is no single history, but rather a sequence of pasts, different ways of expressing a world and oneself. Memory connects the narrator and the character in Proust's narrative voice, while in Swift the relation is an amorphous one, as the narrator is tinted by the character's mind. The narration of *Waterland* is itself a product of a memory embodied in style. The recalled picture is a linguistic image, an image that retains the discursive environment from which it came. In *Waterland,* recollection, which serves also as a source for inner monologue, produces discontinuous articulations. It frequently recalls a discourse fragment that is contextually linked to the present. It circulates bits from other portions of the text that make up the character's past, creating tensions and displacements within the text. Memory in the text's simulacrum is thus a textual memory that replicates parts into new articulations. And with this collision of articulation chains we lose any sense of an identity in command or control.

6 Interiority and Identity through Non-Linearity

In the opening of "De la Révolution," in a passage which can be seen as symptomatic of the novel's narrative encounter with time, Tom contemplates the experience of time and its relation to history.[36] In the passage at the end of the previous chapter, Tom's questioning himself, "And what am I going to do?" functions as a transition, opening the novel up to the world of interior experience. Tom muses, 'how does a man move? One step forward, one step back (and sometimes one step to the side). ... We believe we are going forward, towards the oasis of Utopia. But how do we know ... that we are not moving in a great circle?' The articulation of this world obeys a different logic than that of the objective narrative, which views events and objects from outside ("the great so-called forward movements of civilization whether moral or technological,

35 Henri Bergson, *Matter and Memory*, Cosimo Classics Series (1896; repr., New York: Cosimo Classics, 2007), 71.

36 Swift, 186–89.

have invariably brought with them an accompanying regression"). Here, "a step forward" (an action in everyday life) leads to "moving forward" (an indication of what is yet to come) which turns us towards the discourse of progress and development. The connections between the sentences are associational, as the word "forward," for example, is first part of the phrase "a step forward" and then, disengaged from that context, is made a topic of general philosophical reflection in the following sentence ("forward movement of civilization").

The section not only relates the events, but also grounds them by establishing internal links, creating a contrast between inner monologue and narration. The pieces are arranged to illustrate the progression of a linear sequence of events. Concrete details are provided but do not significantly hamper the narration's development. This contrast between inner monologue and external narration has significant implications for our understanding of time. The outside viewpoint transforms the reader into an observer who understands the links between the numerous events he or she witnesses. As we read further, the separate components fit into place, and the emergent passage supplies us with a continuous and comprehensible totality of temporal experience. The reader, however, also remains unaware of a crucial temporal difference between the time of narration and the time of what is portrayed.[37] As we are transformed into an observer of the internal world, we must also supply the components that maintain coherence in that world. This clearly resists any simple correlation between the time-experience of the monologue and the inferred time-experience of the reader's appropriation of the monologue. But the problem goes considerably deeper. The reconstructed whole does not necessitate a linear horizontal structuring of pieces. Rather, Tom's narration moves in a circular fashion, as well as in divergent horizontal and vertical sequences. This results in a lack of temporal continuity between the external narration and the inside monologue. There appears to be a dispersion or dilation of temporal experience in the monologues. In contrast to the organised linearity of the narration, we have a flow of discrete items related associationally but not integrated into a linear whole.

There is, however, another component to the issues highlighted by Swift's textual organisation, and one where his endeavour parallels that of Bergson. In Bergson's writings, time is seen as durée, "a pure heterogeneity that differs from itself all the time."[38] Durée is the primordial experience of time, a differentiation eventually arrested by the "spatializing and quantifying impulse

37 Henri Bergson, *Time and Free Will: An Essay on the Immediate Data of Consciousness* (1889; repr., New York: Dover Publications, 2012), 44.

38 Bergson, *Time and Free Will*, 17.

in reason" which leads to representation.[39] In generating the proliferation of contrasts that both precede and surpass representation it is pure temporality. In Swift these reflections take on a further linguistic twist. The temporality of the text emerges through patterns of partial syntheses that do not fit into one another, producing instead additional disparities, differences of a second degree. We have only differentiation mechanisms at work, machines that create an increasing number of levels, and modes in which these differences act.[40]

The recurrence of these verbal fragments, however, build into their own contexts. Memory in *Waterland* is made up of a play of contrasts occuring between the former use and the current setting. Memory here does not represent an organic whole with a fixed meaning, but rather a sequence of repeats, each of which leads to new series. An element that was coupled with other components in a previous episode carries traces of these combinations with it later in the novel. Thus, in internal monologue, recall is primarily explored through repetition. As I've suggested, these acts of repetition can transcend the limitations of character-subjectivities and build patterns that go beyond psychological connections, revealing the limitations of an explanation based on stream of consciousness. There is a greater resemblance to Bergson's idea of durée, which Deleuze argues is inextricably linked to the movement of actualization. As he writes, "actualization occurs through differentiation, through divergent lines, and generates so many differences in kind by virtue of its own movement."[41] Similarly, as Udaya Kumar suggests "The profound relationship between durée and non-linearity is based on the continuous differentiation process that informs them."[42] This difference appears in Swift's employment of multiple discourses and repetition. Each recalled linguistic fragment causes a difference, an interaction between its previous and current discursive links. The result is a literary memory within the novel that extends beyond individual characters' perceptions of time. We've seen how repetition disrupts the boundaries between narration and monologue, and between the presentation of subjective and objective time experiences. These occurrences of repetition disrupt linear time as well, making intricate recapitulations necessary to the story. This recursive arrangement, however, does not result in a totality; rather each piece remains available for additional recurrence, a fact Swift emphasises

---

39 Gilles Deleuze, Hugh Tomlinson, and B. Habberjam, *Bergsonism* (New York: Zone Books, 1988), 29.

40 K. A. Pearson and J. Ó. Maoilearca, *Henri Bergson: Key Writings*, Athlone Contemporary European Thinkers (London: Bloomsbury Publishing, 2002), 57.

41 Deleuze, Tomlinson, and Habberjam, *Bergsonism*, 43.

42 Kumar, 79–80.

through the circularity of the novel. We get series by repetition—expanding, differentiating from one another, interacting with one another. This results in various temporal syntheses that cannot be totalized further. Such a process defies both linear and cyclical theories of time.

7 Non-Linear Narration as a Measure of Time

Waterland deconstructs traditional concepts of linear development by exploring the flexibility of time and the interplay between repetition and diversity. To begin, the novel's story structure emphasises the notion of recurrence and variation in measuring time by allowing its plot to shift back and forth between time eras, blurring the line between past and present. This non-linear structure disrupts conventional frameworks of time, positing time itself as an intricate interweaving of repeating events and shifting viewpoints, rather than an unchanging, unidirectional entity. Second, the protagonist, Tom Crick, constantly dwells on and relates his previous experiences in a reenactment of memories and story that reveals the ever-changing nature of time. There are, moreover, slight variances in Tom's recollections, suggesting that the meaning and significance of events are continually susceptible to interpretation and reinterpretation. As Tom recalls the same tales and anecdotes about his forefathers throughout the novel, but adding fresh facts or beliefs with each reiteration, a sense of difference emerges, as the past is constantly recontextualized. This movement reflects Deleuze's idea that repetition is not simply mechanical reproduction, but rather a process of creating new connections and meanings through subtle variations.

The theme of repetition and difference is also evident in the overall structure of the novel. Swift employs a non-linear narrative style, moving back and forth between different time periods and perspectives. This repetition of events from various angles highlights differences in perception and interpretation, challenging the notion of a single, fixed truth. Furthermore, the characters in *Waterland* themselves embody the concept of repetition and difference. Tom's father, for instance, repeats the same routines and rituals day after day, finding comfort in the familiar. However, even within this repetition, there are subtle differences in his actions and behaviours, suggesting that change inheres in repetition itself. The river and the Fens, the novel's setting, also serve as symbols of repetition and difference. The river flows continuously, repeating its course over time, yet unique in each moment, as the water is never the same twice. It thus recalls Deleuze's idea of repeated elements always imbued with difference.

8 Repetition towards Conclusion

The theme of repetition and difference is also explored in the context of memory and personal identity. Tom's recollections of his past are not fixed and stable, but rather fluid and continually changing. Each time he remembers an event, it is influenced by his current perspective and emotions, creating a different version of the past. Additionally, the concept of repetition and difference is tied to the theme of storytelling in the novel. Tom is constantly telling and retelling stories, weaving together different narratives and perspectives. This reflects Deleuze's idea that repetition is not simply a reproduction of the same, but rather a creative act of making connections and generating new meanings. In *Waterland*, repetition and difference are also linked to the theme of history and its relationship to the present. The novel explores how the past repeats itself in various ways, with patterns and echoes reverberating through time. However, each repetition is never exactly the same, as it is influenced by the specific circumstances and contexts of its present. This theme is also closely tied to the measurement of time. As a history teacher, Tom presents historical events as repetitive patterns, highlighting the cyclical nature of history and suggesting again that time does not progress linearly but rather repeats itself in continually different forms.

The Fens, the setting of the novel, also plays a significant role in the measurement of time. The landscape itself, with its vast expanses and slow-moving rivers, appears timeless. The repetition of natural elements, such as the ebb and flow of the tides or the cycle of the seasons, emphasizes the cyclical nature of time and the presence of difference within each repetition. The setting also reflects the way in which the characters' experiences of time reenact the Deleuzian concept of repetition and difference. Tom Crick, for instance, feels trapped in a repetitive cycle, teaching the same lessons year after year. However, within this repetition, there are subtle differences in his approach and perspective, suggesting that time is not stagnant but always evolving.

Finally, the theme of repetition and difference appears again in the characters' aging. As time passes, the characters physically change, and yet preserve echoes of their younger selves in their present state. Again, time appears not as a linear march towards an ultimate endpoint, but rather as a continuous interplay between repetition and difference, where past and present coexist. The theme of memory is also intricately linked to the measurement of time in *Waterland*. Memories are repeated and revisited, but with each repetition they acquire new meanings and interpretations. The act of remembering is not a simple replication of the past, but rather a creative process of constructing and reconstructing the self through an interplay of repetition and difference.

Like the river, time is fluid and ever-changing, defying both definition and measurement.

Overall, Graham Swift's *Waterland* exemplifies the Deleuzian concept of repetition and difference through its exploration of memory and history. The novel reveals that repetition is never a mere reproduction of the same, but rather a creative process constantly generating new meanings and connections through variations and differences. This idea also challenges the notion of time as fitting into a fixed and linear narrative, as the novel disrupts traditional notions of chronology by presenting events out of order and emphasizing the interconnections between different time periods. The approach suggests that time is not a straight line, but rather a web of repeated moments and diverging paths. Ultimately, the Deleuzian concept of repetition and difference highlights the fluidity and multiplicity of temporal experiences in *Waterland*. The novel thus suggests that time is not an external force acting upon individuals but rather an internal and subjective construction, shaped by the interplay of repetition and difference.

References

Ansell-Pearson, Keith. *Germinal Life: The Difference and Repetition of Deleuze*. Taylor & Francis, 1999.

Ansell-Pearson, Keith and John Ó. Maoilearca. *Henri Bergson: Key Writings*. Athlone Contemporary European Thinkers. London: Bloomsbury Publishing, 2002.

Bakhtin, M. M. *The Dialogic Imagination: Four Essays*. Edited by Michael Holquist. Translated by Caryl Emerson and Michael Holquist. 1982. Reprint, Austin: University of Texas Press, 2010.

Bergson, Henri. *Matter and Memory*. 1896. Reprint, New York: Cosimo Classics, 2007.

Bergson, Henri. *Time and Free Will: An Essay on the Immediate Data of Consciousness*. 1889. Reprint, New York: Dover Publications, 2012.

Currie, Mark. "The Novel and the Moving Now." NOVEL: *A Forum on Fiction* 42, no. 2 (2009): 318–25.

Currie, Mark. *About Time*. Edinburgh: Edinburgh University Press, 2010.

Deleuze, Gilles. *Difference and Repetition*. Translated by Paul Patton. 1994. Reprint, New York: Columbia University Press, 2004.

Deleuze, Gilles, Hugh Tomlinson, and Barbara Habberjam. *Bergsonism*. New York: Zone Books, 1988.

Ingold, Tim. "The Temporality of the Landscape." *World Archaeology* 25, no. 2 (1993): 152–74.

Jameson, Fredric. "The End of Temporality." *Critical Inquiry* 29, no. 4 (2003): 695–718.

Kumar, Udaya. *The Joycean Labyrinth: Repetition, Time, and Tradition in Ulysses*. Oxford: Clarendon Press, 1991.
Marker, Chris, dir. *Sans Soleil*. Documentary Film. Paris: Argos Films, 1983.
Rovelli, Carlo. *The Order of Time*. New York: Riverhead Books, 2018.
Smith, Jennifer J. "Writing Time in Metaphors." In *The American Short Story Cycle*, 60–86. Edinburgh: Edinburgh University Press, 2018.
Swift, Graham. *Waterland*. London: Simon & Schuster, 1983.
West-Pavlov, Russel. *Temporalities*. The New Critical Idiom. Oxford: Taylor & Francis, 2012.

CHAPTER 13

Branching Timelines and Measuring the Unled Life in Ted Chiang's "Anxiety Is the Dizziness of Freedom"

Jo Alyson Parker

Abstract

Focusing primarily on Ted Chiang's 2019 speculative-fiction novella "Anxiety Is the Dizziness of Freedom," I address the literary question of parallel worlds and branching timelines. I explain Chiang's indebtedness to the many-worlds theory of quantum physics in conceiving the "Plaga interworld signaling mechanism" (or "prism") that drives the plot. When activated, the prism splits one world into two, with two timelines that immediately begin to diverge. The communication between the branches allowed by the prism facilitates measurements and evaluations of noticeable differences brought about by the passage of time. After examining the various situations brought about by the prism technology, I address the psychological and philosophical implications of "unled lives." I conclude by explaining how Chiang's novella enables us to grasp the psychic costs of speculations on the might-have-been, such as fear of missing out, schadenfreude, and existential angst. As I argue, however, Chiang ultimately shows that examining the unled life can lead to acceptance of and satisfaction with the life one has to live.

Keywords

Ted Chiang – "Anxiety Is the Dizziness of Freedom" – many worlds theory – branching timelines – quantum physics – unled lives

•••

> From time's fluid movement one moment is chosen as the one that made all the difference.
>
> ANDREW H. MILLER, *On Not Being Someone Else*

 | DOI:10.1163/9789004723634_015

•••

Foundation laid by every
act and omission.
each decision, even
those the mind would
label "big mistake."

DANNA FAULDS, "Foundation Stones"

•••

Maybe in another life you get
to live out all the lives you've imagined. Maybe in this life
I become who I am by not knowing—

TIANA CLARK, "Maybe in Another Life"

∴

Deriving primarily from the many-worlds theory of quantum physics, the idea of branching timelines has been a staple of narrative fictions, films, and television series, tantalizing readers and viewers with the idea of the "what if," the unled life that they might have led had they turned left instead of right or gotten into a different elevator or taken a different train—or been subject to the myriad indeterminate and indeterminable factors that, over time, would lead "two branches" of possible experience to diverge "visibly on a worldwide scale," (Chiang 2019a, 285).[1] Often in what we might call parallel-worlds fiction, when a character in one world is made aware of another, ostensibly familiar world, the visible divergences can be enormous (for example, a "good" character in one world may be a "bad" character in another), but in these cases, the causes for the divergence are seldom addressed.[2] In his provocative and provocatively

1 See Front (2015, 37–38) for a discussion of the various fields in which the idea of many worlds appears. The term "branching" is prevalent in discussions of many-worlds theory—and in Chiang's story—but it suggests a primary world out of which various branches grow. The more accurate term may be "splitting," which suggests one world dividing into two. I use "split" to designate what happens when parallel worlds are generated, but, following Chiang, I use "branches" to designate the worlds that have been generated.

2 I think particularly of the *Star Trek: TOS* episode "Mirror, Mirror" (1967), wherein denizens of the benevolent Federation are transported into a parallel universe that is the Federation's evil counterpart. This parallel universe continues to crop up in *Star Trek* spin-off

titled novella "Anxiety Is the Dizziness of Freedom," however, Ted Chiang focuses on the initial conditions that influence divergences and advances the idea that people within the various branches can communicate for a limited time, thus enabling his characters to evaluate—that is, measure—the overall differences in their now-separate lives brought about by the passage of time.

The device that enables Chiang's characters to communicate with selves in alternate timelines, or "paraselves," is the "Plaga interworld signaling mechanism," familiarly called a "prism" (Chiang 2019, 272, 273).[3] In developing this fantastical device, Chiang alternates throughout the novella between two diverse genres: (1) pseudo-historical, objectively recounted sections that explain the mechanics, implementation, and effects of the prism and that reinforce the realism of the narrative; and (2) narrative sections that focus on the adventures of individual prism-users and that bring to life the explanations contained in the objective segments. This alternating structure also enables Chiang to bypass the often awkward moments of exposition that sometimes occur in speculative fiction.

Throughout the novella, Chiang explores both the harms and benefits of examining lifepaths not taken. As Chiang's characters confront their paraselves' lives, they may experience a fear of missing out, schadenfreude, and even the existential angst generated by a proliferation of life choices. Yet Chiang also explores how examining the unled life may lead his characters to self-acceptance and self-actualization. Although we do not have paraselves against whom we can compare ourselves, Chiang's realistic fantasy prompts us as well to imagine the unled lives that we might have lived. In so doing, we may experience feelings similar to those of Chiang's characters. Chiang foregrounds these issues through a narrative that measures the contrast between branching timelines.[4]

---

series, including the recent *Discovery*. A similar distinction occurs in the short-lived series *Counterpart* (2017–2019), which features a milquetoast office worker in one universe who has become a ruthless agent in the other, both characters played by J. K. Simmons. *Counterpart*, however, does put forward an origin story for the diverging universe (an experiment in Berlin 30 years ago that went awry and caused the rift) and does suggest a possible reason for the diverging personalities (a devastating pandemic in one of the worlds).

3 Chiang points out that the name "prism" is "a near acronym of the original designation" (Chiang 2019a, 273). As such, it invokes the glass prisms with which we are familiar. Just as a glass prism refracts many different colors from a single beam of white light, each of Chiang's prisms refracts two different worlds.

4 This article serves as a companion piece to a 2021 article, co-written with Thomas Weissert, that deals with what we call "eternal recursion"—time loops in fictional texts that enable do-overs on the part of one (or even several) characters (Parker and Weissert 2021). As we argue, although such do-overs are impossible in reality, they spur us to consider what might

1 Many-Worlds Theory, Chiang's "Prisms," and Exponential Changes over Time

In the narrative sections, Chiang shifts the point of view between two key characters: (1) Natalie, a woman formerly suffering addiction who now works at a failing business called Selftalk, which rents prisms to people who do not own their own; and (2) Dana, a therapist who runs a support group for people struggling with prism-addiction. Their lives intersect when Natalie, at the instigation of her boss, Morrow, joins the support group to persuade another member, Lyle, to sell his potentially valuable prism to Selftalk. During her time in the support group, Natalie also discovers that Dana is being blackmailed by an old high school friend, Vinessa. When Dana and Vinessa were found in possession of drugs during a high-school trip, Dana shifted all the blame onto Vinessa, an act that Dana sees as the cause of Vinessa's subsequent downward spiral. The adventures of Natalie and Dana, as well as those of some of Natalie's customers and Dana's clients, provide concrete examples of the various effects and outcomes of prism use.

In the pseudo-factual sections of the novella, Chiang draws upon actual science—in this case, the many-worlds theory of quantum physics—to undergird his science-fiction invention, its apparent plausibility both facilitating our suspension of disbelief and enhancing the imaginative suppositions provoked by the story.[5] Many-worlds theory was first put forward by H. Everett III in 1957 and reinforced ten years later by Bryce DeWitt, who introduced "the terminology 'many worlds' and 'many universes' in print" (Saunders 2010, 6). This theory presupposes "an infinity of other parallel worlds, equally existent, equally real, where an infinite number of copies of myself exist and experience every sort of

have been. The parallel worlds and branching timelines that I examine here similarly inspire us to consider the might-have-been.

5 One of the compelling aspects of Chiang's stories is his creation of devices that seem *as if* they might indeed exist in the real world and that enable Chiang to tease out the implications of certain current technologies and trends. In the collection *Stories of Your Life and Others* (2016, 2022), see the mind-enhancing "Hormone K therapy" (from "Understand") and the programmable pharmaceutical "neurostat" that enables the creation of "calliagnosi" to combat "lookism" (from "Liking What You See: A Documentary"). In the collection *Exhalation* (2019b), see the "digients" or digital organisms that live in simulated environments (from "The Lifecycle of Software Objects"), the "Predictor" that reveals free will to be an illusion (from "What's Expected of Us"), and the "lifelogs" that record one's life and the related search tool "Remem" (from "The Truth of Fact, the Truth of Feeling"). I discuss the Predictor in "Ted Chiang's Time Travel Narratives" (2022) and the lifelog and Remem in "Ted Chiang's Braided Stories of Memory, Narrative and Technology" (2021).

alternative reality" (Rovelli 2021, 58).[6] Certain versions of the theory hold that there is a shared "common initial segment," which David Lewis describes in *On the Plurality of Worlds*:

> In branching, worlds are like Siamese twins. There is one initial spatiotemporal segment; it is continued by two different futures [...] and so there are two overlapping worlds. One world consists of the initial segment plus one of its futures; the other world consists of the identical initial segment plus the other future.
>
> 1986, 206

Chiang predicates prism technology on these ideas of a "common initial segment" and "an infinity of parallel worlds." He conceives that each time a prism is initially opened, one world divides into two, binding that particular prism to these two worlds. Initializing a different prism would then generate a new pairing of worlds. In the initial segment, only one self (or character, in Chiang's novella) exists, but once the initial segment splits into two, this one self becomes two separate selves (or characters).[7] Throughout the novella, Chiang gives us instances of various selves whose paraselves inhabit different worlds that have split from one initial segment. The first episode of the novella, in fact, features Natalie communicating with a paraself that is new to her.

We might regard an inactivated prism as analogous to Schrödinger's well-known box in which a cat exists in two distinct states—a superposition of cat alive and cat dead—until the box is opened and an observation, or measurement, can be taken.[8] Carlo Rovelli extrapolates to the many-worlds

6 Although offering a lucid explanation of the theory, Rovelli nonetheless calls it "crazy" (59). Despite such objections, it has provided a great deal of fodder for science-fiction texts. For less skeptical accounts of the theory, see Lewis (1986) and Saunders (2010).

7 Lewis himself dismisses the idea that one initial segment splits into two in favor of what he calls divergence: "In divergence, on the other hand, there is no overlap. Two worlds have two duplicate initial segments, not one that they share in common. [...] Not I, but only some very good counterparts of me, inhabit these other worlds" (1986, 206).

8 The paradox appears originally in Schrödinger 1935 (translated by Trimmer 1980) in which "[a] cat is penned up in a steel chamber" with "a tiny bit of radioactive substance," and "*perhaps* in the course of one hour one of the atoms decays, but, also, with equal probability perhaps none." If the decay occurs, hydrocyanic acid is released. The paradox occurs because we cannot know whether the acid has been released without opening the box and making an observation: "the cat still lives *if* meanwhile no atom has decayed. The first atomic decay would have poisoned it. The ψ-function of the entire system would express this by having the living and the dead cat [...] mixed or smeared out in equal parts" (328; Schrödinger's emphasis).

implications of the paradox: "The world has branched into two: one world in which the cat is awake, and I see the cat awake, and another world in which the cat sleeps, and I see him sleeping. There are now two versions of myself: one for each world" (2021, 57).[9] The prism behaves similarly (with different colored lights substituting for the cats), as Chiang's stimulating pseudo-quantum-theory explanation makes clear:

> When a prism was activated, a quantum measurement was performed inside the device, with two possible outcomes of equal probability: one outcome was indicated by the red LED lighting up, while the other was indicated by the blue one. From that moment forward, the prism allowed information transfer between two branches of the universal wave function.
>
> 274

For example, Lyle, the young man whom Natalie eventually inveigles into selling his prism, bought the prism in order to decide whether he should accept a new job, basing his decision on the outcome of the quantum measurement: "The blue LED had lit up in this branch, so he took the job while his paraself [whose red LED had lit up] stayed at his existing job" (280).[10]

Chiang suggests that, before Lyle opened his prism, his two potential futures were in superposition, but that once he opened it, two separate Lyles (self and paraself) embarked on different paths, based on the different decisions that they had taken.[11]

---

9 An apparent ailurophile, Rovelli prefers a sleeping cat to the original formulation of a dead one.

10 The knowledge that one's paraself plans to make the opposite decision because of the color of the light is an example of "quantum entanglement," whereby two particles are linked across distances. It thus gives each branch definitive knowledge of what is happening in the other branch—at least in the moments after the opening of the prism. I thank Thomas Weissert for this observation.

11 Which is "self" and which is "paraself" is an intriguing question. For instance, I would be the self in, say, branch A, with my counterpart the paraself in branch B, but the reverse holds true as well, the point being that, from the moment the parallel worlds open up, both people regard themselves as *the* self, and neither, from that point on, has access to the other self's inner workings. I am *the* self only by virtue of being in the branch in which I exist, and my paraself is no longer me but an *other* to me, as I would be to my paraself. In his cogent analysis of Larry Niven's classic many-worlds story "All the Myriad Ways" (2001), David Wittenberg notes that parallel-world stories function as "ontological thought experiments […] introducing the disturbing possibility that there is no natural criterion for preferring certain lines over others" (2013, 17). For a comprehensive philosophical and scientific account of the problem of identity across multiple worlds, see Front (2015, 47–64).

A particularly significant feature of the prism technology for the characters who use it—and for Chiang's plot purposes—is "the information transfer between two branches," which enables a self and paraself to communicate and measure the divergences in their lives over time. When Natalie opens the unfamiliar prism in the first episode, she is struck by the fact that her paraself is wearing a necklace she doesn't recognize, a slight but nonetheless observable divergence.[12] Chiang provides a more dramatic example of divergence with Lyle. After opening his prism, Lyle keeps track of his paraself's career trajectory, which improves as Lyle's worsens. (Presumably, Lyle's paraself is also marking this distinction.) Savvy entrepreneurs sell information to subscribers about parallel versions of current events, allowing comparisons between sports teams, celebrity behavior, and so forth. People can even assess "different versions of novels published in different branches, with the result that authors faced competition from pirated copies of books they might have written" (297). (One wonders what a parallel version of Chiang's novella might look like.) From the outset, Chiang sets up the idea that the differences between the diverging paths can be measured over time.

Chiang devises some constraints for his device, thus avoiding a certain gimmickry inherent in some other timeline stories. In keeping with the idea of multiple worlds, Chiang presupposes that the splitting has been occurring all along, before the invention of the prism, and, indeed continues to occur without it, "generated by any quantum event" (326). The prism simply enables heretofore unavailable access to branching timelines. Significantly, however, this access is *only* to the forward-moving branch generated by that particular prism, with the past of each world (the "common initial segment") always remaining the same. Furthermore, there is no access to branches that have occurred before prism technology, nor is there access to future branches that occur due to non-prism-generated quantum events. Chiang elaborates upon this constraint: "There'd be no reports from branches where Kennedy hadn't been assassinated or where the Mongols invaded Western Europe. By the same token, there were no fortunes to be made by patenting those inventions gleaned from branches where technological progress had taken a different route" (294). This constraint enables Chiang to avoid the entertaining but sometimes chaotic motifs that crop up in some classic branching-worlds stories: for example, Bruce Sterling and Lewis Shiner's "Mozart in Mirrorshades" (1985), wherein people have figured out a way to go back into various periods of history and exploit the resources there (stealing classic paintings, drilling for

12 Note that this particular prism is not one that has been initially opened by Natalie, but it is one in which a para-Natalie nevertheless exists. She might not work at Selftalk or might even no longer exist in a parallel world bound to a different person.

oil) in exchange for magazines, cosmetics, chocolate bars, and so forth; or Larry Niven's "All the Myriad Ways" (1989), wherein "the Crosstime Corporation" holds patents for inventions imported from other timelines (101).[13] By avoiding this kind of time-line manipulation, Chiang can keep our focus on the issue of measuring one's life trajectory against that of one's paraself.

Although the prism technology enables what seems to be an opportunity to talk with the dead (some bereaved people are able to get updates of their deceased loved ones' social media accounts from a timeline where they are still alive), this ability is fleeting, for Chiang imposes an additional constraint. Each prism has a limited lifespan, "like a notepad that the two branches shared, and each time a message was sent, a strip of paper was torn off the top sheet," leading eventually to the device being used up "so that no more information could be exchanged and the two branches went their separate ways, incommunicado forever after" (274). Chiang reinforces the sense of loss that might accrue in a subplot involving Natalie. Following her boss Morrow's directives, she hopes to sell the prism she received from Lyle to the celebrity Scott Otsuka, who had recently lost his husband, Roderick Ferris, in a car crash. In the alternative timeline opened up by Lyle's prism, however, Roderick Ferris survived and Scott died. Lyle's prism would thus enable Scott to communicate with para-Roderick. As Natalie bargains with Scott, both Scott and para-Roderick make clear that they are aware of the prisms' ephemerality:

> [T]hey had talked about the finiteness of what the prism offered. No matter how much [Scott] and Roderick tried to conserve, the amount of time left on its pad wouldn't last the rest of their lives. […] the pad

13 Both stories have been collected in Henry Turtledove's 2001 anthology *The Best Alternate History Stories of the 20th Century*. Sterling and Shiner's story inspired William Gibson's *The Peripheral* (2014) and its sequel *Agency* (2020), both of which deal with branching timelines, therein called "stubs," a term that suggests a primary world out of which they grew. In Gibson's novels, as in Chiang's novella, communication takes place among stubs so that, for example, one stub can be made aware that in another stub Hillary Clinton was elected to the U.S. Presidency in 2016 after the stub branched off. There are, in fact, various instances wherein Gibson's characters explore the significance of variations among stubs. Gibson also deals with the issue of prioritizing one world as *the* world. As one character notes while deploring the "regrettable" term "stubs" for other worlds, it is "inaccurate as well, since your continuum won't remain short. It appears so to us, but only since it's diverged from our shared past" (2020, 200). Unlike Chiang, Gibson does not provide a technological explanation for the generation of stubs, and characters themselves are baffled by their mysterious generation. There is, by the way, an entire sub-genre of alternate-timeline stories in which John Kennedy was not assassinated or survived the assassination.

> would eventually run out, and then they'd have to say goodbye. Scott had been willing to go ahead with it; any extra time they had together was worth it, and when the end came, at least it wouldn't come as a surprise.
>
> 331

In this passage, Chiang makes clear that this constraint is fixed and irrevocable. As well as providing a certain poignancy to the encounters a prism may facilitate, Chiang's restriction of finiteness forecloses the possibility of infinite measurement and amplifies the technological plausibility of the prism (computers wear out, batteries die).

As Chiang's example of Scott, Roderick, and their paraselves makes clear, once a prism is opened, each branch begins to evolve separately over time, often leading to great differences. What prompts this separate development? Chiang explains that, in certain cases, the deviation arises from well-defined opposing actions that a singular self has determined before the prism's opening: "If a person has resolved to base a huge decision on the measurement—'If the blue LED lights up, I will detonate this bomb; otherwise, I will disarm it'—then the branches will diverge in an obvious manner" (279). Like Lyle, people may base decisions on the seemingly random circumstance of a light's color, as if they were flipping a coin. In general, however, Chiang eschews such dramatic causes and, in keeping with many-worlds theory, presupposes that the prism's very activation triggers a divergence as infinitesimal as "whether two oxygen molecules collide or whether they drift past each other" (284). Once again, Chiang reinforces the technological plausibility of the fictional device, even in his explanation, which cites renowned physicist Michael Berry, who "calculated that you could predict only nine collisions [of a given billiard ball] before you would need to account for the gravitational effect of a person standing in the room" (284).

To account for the effects of such a seemingly "imperceptible […] discrepancy at the level of the thermal motion of molecules" (284), Chiang, without making an explicit reference, alludes to the sensitive dependence on initial conditions (sometimes called "the butterfly effect") popularized by dynamical systems or chaos theory—that is, the fact that, over time, a small perturbation can have exponentially large effects. Edward Lorenz discerned such a sensitive dependence when he carried out experiments on the weather, which he published in papers now regarded as the antecedents of chaos theory.[14] In a possible homage to Lorenz, Chiang turns to the weather when describing

---

14 James Gleick offers a compelling narrative of Lorenz's discovery in his chapter "The Butterfly Effect" in *Chaos: Making a New Science* (1987, 9–31).

the divergences between prism branches: "The growth of error in weather prediction is identical to the divergence between the weather in the branches on the opposite side of a prism. The initial perturbation is the difference in the collision of oxygen molecules when the prism is activated, and within a month, the weather around the globe is different" (285). Chiang even invents a Lorenz-like scientist named Peter Silitonga, whose initial forays into "prism physics" involve activating a prism and one month later exchanging weather reports with his paraself in the new timeline, thereby discovering that "the two branches had diverged visibly on a worldwide scale" (285)—in effect, measuring the divergences between branches.[15] Silitonga and yet another paraself even demonstrate that if children are conceived by a once-identical set of parents in the two branches, they will end up with "a different chromosomal makeup than their counterpart in the other branch," thus establishing that "the smallest change imaginable will have global repercussions" (290). The (fictional) researchers find that prisms enable them "to study the mechanisms of historical change" and "to monitor the news over the following weeks and months *to see how the divergences grew over time*" (my emphasis). In a classic case of chaotic effects, however, they discover "small discrepancies unrelated to the one they'd originally discovered; the weather was instigating changes everywhere, all the time" (295–96). Chiang's examples highlight a key idea in many-worlds theory—that parallel worlds are "not spatially, but *dynamically* separated" (Saunders 2010, 5; my emphasis)—that is, that their separation is characterized by changes over time.

Chiang conceives that Silitonga and his paraselves' experiments (such as measuring divergent weather patterns or divergent chromosomal patterns in babies) are, like those of actual scientists, dependent upon a control, a method of improving the likelihood of a meaningful measurement. These cases involve each scientist starting from exactly the same initial condition, such as identical sets of parents or the weather conditions prior to the opening of a prism. As I have suggested elsewhere, Chiang's fantastical devices enable him to perform thought experiments, that is, to consider the implications of "what if"—what if we had a time machine, a lifelog that recorded everything that we did from birth to death, an implant that keeps us from noticing the distinction among people's looks.[16] For plot purposes, Chiang, like the experimentalists within his

15 The resulting academic paper—"Studying Atmospheric Upscale Error Propagation with the Plaga Interworld-Signaling Mechanism"—may be Chiang's sly jab at academese.

16 See Parker 2022, 138. The time machine occurs in "The Merchant and the Alchemist's Gate" and the Lifelog in "The Truth of Fact, the Truth of Feeling" (*Exhalation* 2019b). The callignosia implant appears in "Liking What You See: A Documentary" (*Stories of Your Life* 2010).

story, devises what we might regard as a control to demonstrate the psychological and philosophical repercussions of measuring one's actual life against an unled life (whether that of a paraself or an imaginative projection). In the novella, this control is the idea of behavioral consistency between a character and their paraself—the idea that one would act in a largely consistent manner across multiple timelines, with divergences between them arising from different external circumstances. Indeed, Chiang elaborates upon this idea in his "Story Notes," in which he addresses the idea of behavioral consistency across worlds (or life circumstances) by using Martin Luther as an exemplar: "If you could somehow examine a multitude of Martin Luthers across many worlds, I think you'd have to go far afield to find one that didn't defy the church, and that would say something about the kind of person he was" (2019c, 349). According to Chiang, self and paraself not only start from the same initial conditions with regard to behavior but also maintain this behavioral consistency across worlds. Only thus, Chiang suggests, can they make meaningful measurements of how small changes generated by opening the prism may lead to global divergences in their lives.

2 The Road Not Taken and Measuring Choices Made over Time

Chiang demonstrates that his characters' confrontation with the diverging paths taken by their paraselves may lead to both harmful and beneficial effects, and that it is useless for the characters to anticipate which will occur. The first, and possibly most pernicious, negative effect of measuring a self against a paraself is the fear of missing out—a malady so prevalent in our own society that it has been granted the acronym FOMO.[17] Chiang shows how Lyle is in thrall to FOMO, regularly contacting his paraself via his prism and fretting that the crucial decision he made when he first opened it has led him down the wrong path. Although, initially, both Lyle and his paraself had been satisfied with their opposing decisions, over time the branched lives have evolved differently, so that Lyle ends up "disenchanted with his new duties," especially when his paraself gets a promotion (280). When his paraself embarks on a successful relationship with a woman named Becca, an envious Lyle contacts the Becca who exists in his own branch, sends her a photo of his paraself with

17 In 2004, Harvard MBA student Patrick McGinnis coined the term "FOMO" in a humor column for *The Harbus* (2004). As he later wrote, he didn't realize until he was contacted for an interview in 2014 that his "word had lived an exciting life of its own." After the "article went viral," McGinnis went on "write a book, do a TED Talk, and create a podcast, all about how FOMO shapes decision-making in business and life" (2014).

para-Becca, and persuades her to go on a date that, perhaps not surprisingly, ends disastrously. This outcome enhances the feelings of jealousy that Lyle already directs at his paraself—feelings that ultimately compel Lyle to give up his prism. As a fellow member of the support group explains, "when we see our paraselves experiencing good fortune, we think we're entitled to the same good fortune" (282). Through Lyle's experiences, Chiang underscores this notion: Lyle thinks he is missing out on the life he should have had—a promotion at work, a compatible relationship. A comparison with a seemingly more fulfilling parallel life can lead to unhappiness.

In describing Lyle's feelings of FOMO, Chiang taps into what has become an increasingly common concern in our own society. Although we do not have paraselves against whom we can compare ourselves, we can and do consider our unled lives—those "moments when, for good or ill we feel confronted by our unrealized possibilities," in Joshua Rothman's words (2019). What if we had majored in biology and gone to med school? What if we had not gone to that crucial dinner party? What if we had accepted that job in Illinois? Would we have been richer, happier, more fulfilled? In *On Not Being Someone Else*, Andrew Miller notes that dissatisfaction may result from such musings: "With a slight settling of my mood, my imagined life makes this one seem like less. Instead of adding to the world, my unled life subtracts from it" (2020, 49).[18] Certainly, with his portrayal of the dissatisfied Lyle, Chiang suggests as much regarding the unled life. We might wonder if Chiang's make-believe prism is in some ways analogous to the real-world electronic devices that make available continuous connections to social media sites such Facebook, Instagram, TikTok, et cetera. As has been often stated, perusal of such sites exacerbates feelings of FOMO in our society, especially when we see someone who may have started on a similar track as ourselves succeeding where we have failed. We can see pictures of former high-school acquaintances enjoying lavish vacations, or we can google a former grad-school rival and discover that they now have an endowed chair at a prestigious university. In his novella, Chiang suggests that examining an unled life may generate feelings that one's own life has fallen short.

A second negative effect of prism use, according to Chiang, is schadenfreude, the psychological quirk of deriving pleasure from someone else's misfortune. One of Dana's other clients, Teresa, had turned down a marriage proposal five

18 Psychoanalyst Adam Phillips makes a similar point: "The myth of our potential can make of our lives a perpetual falling short […]" (2017, xiii). I am indebted to Rothman's essay for drawing my attention to Miller's and Phillips's studies, as well as Kieran Setinya's philosophical treatment of unled lives.

years back but now wonders whether her decision was the right one. She has thus visited a "data broker" who may be able to locate a para-Teresa in another timeline who had accepted the proposal. When Dana asks Teresa what answer she would want if she could interview the paraself who had done so, Teresa makes clear that she wants the paraself's decision to have been a bad one: "I guess what I'd like to find out is that a version of me married Andrew and then divorced him because he wasn't the right guy for me. What I'm afraid of finding out is that a version of me married him and is now blissfully happy" (277). Rather than be happy for a para-Teresa's possible good fortune, Teresa hopes that her paraself's life may be worse than her own—an ironic touch given that the paraself shares an identity with the self. (We might also speculate that para-Theresa harbors similar feelings.) As Dana points out, "It could be that nothing you find out about some other branch will change your situation here in this branch" (277). Teresa's goal, however, does not seem to be bettering her own life but deriving satisfaction from another's misfortune. Chiang prompts us to consider whether we, too, would hope that our imaginary paraselves have lives less satisfying than our own. More pertinently, he asks us to consider how we often make use of something analogous to a "data broker"—for example, googling former loves or professional rivals—in hopes that we might derive satisfaction from seeing what appears to be their less happy or less prosperous lives.

In addition to exploring the FOMO and schadenfreude prompted by encounters with branching universes, Chiang also explores the existential anguish that might occur when one is confronted with a proliferation of possible choices, a proliferation that the existence of multiple worlds highlights. He articulates this plight through the disgruntled son of a woman swindled by Natalie's boss, Morrow. After he pulls a gun on Morrow, who refuses to reimburse him, the son notes this irony: "You work with prisms. You know there's some timeline where I shoot you right now." Although Morrow points out that if the son in this timeline commits the crime, he would be the person going to jail, the son ultimately embraces what seems to be an existential nihilism: "'Fuck it,' he said, 'what difference does it make?' He raised the pistol, shot Morrow in the face, and walked away" (319). Here Chiang may be alluding to Larry Niven's classic branching-timeline story "All the Myriad Ways," wherein "[e]very decision you made both ways. For every wise choice you bled your heart out over, you made all the other choices too." As the key character, Gene Trimble, wonders, "If every choice was canceled elsewhere, why make a decision at all?" ([1989] 2001; 105). In Niven's story, the meaninglessness of decisions entails a proliferation of suicides and murders: "Casual murder, casual suicide, casual crime. If alternate universes are a reality, then cause and effect are an illusion. [...] You can do anything, and one of you will, or did" (106). Niven envisions a chaotic world (or multitudes of worlds), wherein wealthy men plunge off their balconies and wherein Trimble ultimately picks up a gun and shoots himself. (Or perhaps he does—different scenarios play out on the final page.) Certainly, Morrow's killer seems to have embraced this philosophy, as the story illustrates.[19]

Chiang returns to the issue of existential angst when, later in the story, Natalie attends one last meeting of Dana's support group and asks whether there was always going to be some version of Morrow's killer who would pull the trigger. In fact, she ponders whether her decisions matter when she might make opposing choices in a parallel universe: "But when I have a choice to do the right thing or the wrong thing, am I always choosing to do both in different branches? Why should I bother being nice to other people, if every time I'm also being a dick to them?" (327). Through Natalie's question, Chiang reinforces the implications of part of his title—"the Dizziness of Freedom," which engenders anxiety when a person is confronted with the possibility of acting in "all the myriad ways" (to draw on Niven's title as well) that one might act. He indicates that the very idea of a proliferation of choices, particularly highlighted in many-worlds fictions, can strip a person of the comforting illusion that they are following a set path through life.

As with the other issues caused by prism technology, this issue, Chiang implies, also pertains to us readers. Max Axvig (2019) points out that Chiang's title comes directly from a passage in Søren Kierkegaard's *The Concept of Anxiety* in which one's confrontation with a multitude of choices leads to a dizzying anxiety. We might consider, too, Jean-Paul Sartre's extrapolation from Kierkegaard and Martin Heidegger as he discusses the various possibilities he would face when, standing on a precipice, he is confronted with an abyss into which he might fall: "Vertigo announces itself through fear; I am

19 In his analysis of Niven's story, Wittenberg notes that the idea that "an infinity of moral choices is equivalent to no choice at all" is "a rather pallid psychological insight" (2013, 20)—although he also addresses Niven's more complicated take on the issue. For a more recent example of this nihilistic philosophy, consider Lev Zubov (played by JJ Field), one of the masterminds in the television series *The Peripheral* (2022). In the episode "Jackpot," he claims that he has managed to kill off versions of himself and his family in another timeline (an earlier one, according to the set-up of the series), committing a strange sort of murder/(para)suicide. As he explains when referring to multiple universes, "the possibilities are essentially infinite. But one thing troubled me, though. The idea of a different version of myself in another world. The existential nihilism of it" ("Jackpot" 2022). There is no equivalent passage in William Gibson's novel *The Peripheral*, upon which the series is based.

on a narrow path—without a guard-rail—which goes along a precipice" (1956, 30). Fear arises from external causes, such as slipping on a stone and falling. But there is also the possibility that Sartre might *choose* to throw himself into the abyss—that there are no extrinsic guard-rails that would keep him from self-destruction.[20] Although we do not confront other versions of ourselves in parallel worlds, we may imagine other versions of ourselves who ignore the guard-rails of expected behavior as we consider the *what if* in our own lives. Chiang's novella opens us to that possibility.

However, despite his elaboration of the potential consequences of prism use, overall Chiang emphasizes the beneficent effects of examining unled lives, whether in a parallel universe or an imaginative projection. Despite the discontented Lyle and Teresa, the majority of people who use the prisms in Chiang's novella are satisfied with the life that they are leading. Like Teresa, characters who want "to learn about the other paths their life might have taken" (300–1) consult data brokers whose stockpiled prisms have been activated long enough so that "the divergences had grown large enough to be interesting," enabling customers to measure over time what have possibly become large divergences in their successes and failures (300–1). The results are telling: "Predictably, some individuals became depressed after learning that their parallel selves had enjoyed successes that they themselves hadn't. [...] However, most people decided that they liked more things about their life than they did about their parallel selves' lives, and so concluded that they had made the right decisions" (301). According to Chiang, although his characters' attitude may simply be an instance of "confirmation bias" (301), what he proposes in this passage is that people in a prism-prevalent world recognize how their decisions have led to who they are and what they have today. Different decisions would have led to a different life entirely, as the prism gives them the opportunity to see.

The passage speaks to us readers as well. As Danna Faulds remarks in the epigraph at the outset of this piece, "every act and omission" forms the foundation of who we are (2002). In accordance with the butterfly effect, even small alternative decisions in our past could cancel out major facets of our present lives. As Rothman poignantly observes, "If I'd done things just a little bit differently, my son might not exist. Perhaps, in a different life, I'd have a different

20 The passage from Sartre continues: "But I am in anguish precisely because any conduct on my part is only *possible*, and this means that while constituting a totality of motives *for* pushing away that situation, I at the same moment apprehend these motives as not sufficiently effective. At the very moment when I apprehend my being as *horror* of the precipice, I am conscious of that horror as not *determinant* in relation to my possible conduct" (1956, 31; Sartre's emphasis).

wife and child. But I love these particular people; I don't want alternative ones" (2019).[21] Of course, we can only speculate about what our unled lives would look like, but such speculations should probably include not only what we *might* gain but also what we *would* lose. Although we may never divest ourselves of the feelings of FOMO and schadenfreude that beset Lyle and Teresa, Chiang prompts us to consider what we like about the life that we do lead.

Furthermore, Chiang puts forward an alternative to the nihilistic philosophy expressed in Niven's "All the Myriad Ways" and by Morrow's killer. The idea of consistency of character functions not only as a control, enabling the measurement of one lifepath against another, but also as Chiang's rejoinder to the idea that one might, even will, "do anything," in Niven's words. When Natalie questions whether the decision she makes in her world is canceled by a decision a paraself makes in a parallel world, Dana responds, "I think your actions are consistent with your character. [...] If you're someone who's always loved animals, there isn't a branch where you kick a puppy just because it barked at you" (327).[22] Through Dana, Chiang reinforces the point that he made in his "Story Notes" about a multitude of similarly behaved Martin Luthers, and, in so doing, he suggests that there are, in fact, guard-rails that might keep his characters from launching themselves into the abyss.

Chiang further reinforces the idea of behavioral consistency as a comforting constant in two key episodes in the novella. Another of Dana's clients, Jorge, fretting over what he regards as uncharacteristic behavior (puncturing the tires of his mean-spirited manager's car), goes to a data broker to find out whether six other paraselves had behaved differently (they had), in effect taking measurements of their behavior as compared to his own. Although he does not think that his malicious behavior was canceled out by what his paraselves did

21 Henry Turtledove makes a similar point in his introduction to *The Best Alternate History Stories of the Twentieth Century*, explaining how his reading of L. Sprague de Camp's *Lest Darkness Fall* led to his getting a particular college degree, becoming a writer, marrying, and having children (2001, vii)—a seemingly minor experience that led to major life events.

22 Dana's statement regarding consistency of character implies an essentialism with regard to selves and paraselves. Interestingly, this idea seems to bypass the idea of a sensitive dependence on initial conditions that is integral to Chiang's plot. If, indeed, from the moment of prism activation, a self and a paraself are already on diverging life trajectories (taking a new job or staying with the old one), new memories and new knowledge will begin to accrue, possibly leading to very different behaviors. Seemingly, one of the selves could start having many bad experiences with dogs and might eventually end up kicking a puppy that barked at them. The television series *Counterpart* presents a different take on the question of behavioral consistency with the milquetoast office worker and the ruthless agent demonstrating a radical divergence in character over a 30-year period.

not do, he regards his paraselves' behavior as "an indicator of the type of person I am" (287)—that is, according to Chiang, a person who is inherently decent. More significant to the novella's plot is what Dana discovers when she receives a package from an anonymous sender that contains a tablet with video files featuring parallel versions of herself discussing the different actions they performed when they and the Vinessa paraselves were young and confronted with having drugs. One Dana paraself took all the blame upon herself, one said that both she and Vinessa were to blame, and other paraselves reported other choices. In each timeline, Dana discovers that para-Vinessas ended up "stuck in patterns of self-destructive behavior" no matter who was blamed for the drug incident, and she realizes that she no longer needs to "make amends for what she'd done, trying to fix Vinessa's life" (333–39). In both episodes, Chiang shows his characters accepting themselves as they notice behavioral consistency across branching timelines.

With Natalie, Chiang takes the issue of consistent behavior a step further, suggesting that repeating a certain kind of behavior over time can lead to positive self-actualization. As Dana advises her, "Each time you do something generous, you're shaping yourself into someone who's more likely to be generous next time," and so "ensuring that more and more of the branches that split off from this point forward are populated by better versions of you" (329). Thus, in the final episode featuring Natalie, when she attends a Narcotics Anonymous meeting, she talks of doing something nice for someone and adds, "I imagined what a better person might do, and I did that instead" (336), thus measuring herself against a theoretical paraself ideal. She is consciously propagating through time a version of herself that might be consistently good, this behavior thus suggesting a hopeful alternative to the existential angst that proliferating choices may present.

To address how this idea of behavioral consistency might resonate for Chiang's readers (as well as for his characters), I return to Chiang's Story Notes and his Martin Luther example. Chiang argues against those who might claim that the many-worlds interpretation "renders our decisions meaningless, because, whatever you do there's always another universe in which you make the opposite choice, negating the moral weight of your decision" (2019c, 349). Instead, Chiang posits not only that "an individual's character is revealed by the choices they make over time" but that "an individual's character would also be revealed by the choices they make across many worlds" (349–50). Thus, according to Chiang, one's choices have moral weight—and he seems here to speak both about the fictional characters who can access a fantastical prism introducing them to other selves in other worlds, and to readers who weigh different choices that they might have made in their imagined unled lives.

Just as time-loop fiction enables readers to envision a do-over in their own lives, branching-timeline fiction enables them to envision a do-in-addition, an opportunity to consider what kind of life an alternative version of themselves might be leading if they had made a different decision at a crucial bifurcation point. Chiang's contributions to the branching-timelines subgenre center on his ability simultaneously to ground his fantastical inventions in the language and thinking of the actual science—in effect, to make the fantastical plausible—and to create scenarios wherein his characters engage in eminently human behavior. In *Missing Out: In Praise of the Unled Life*, Adam Phillips asks, "The unexamined life is surely worth living, but is the unlived life worth examining?" (xi). Through his elegant fusion of the imaginary and the realistic, Ted Chiang encourages readers to consider why it might, in fact, be.

3 A Coda

Although I have argued that Chiang's story of selves and paraselves encourages us to consider the implications of examining unled lives, I want to return to the fact that Chiang is, after all, a speculative fiction writer, interested in invoking speculations about the quantum physics that undergirds his fantastical prism. As I have suggested earlier, Chiang's prism is analogous to Schrödinger's box, and so Chiang brings the novella to an end by figuratively invoking that box through three episodes. When Natalie is completing final arrangements for selling the valuable prism to Scott, she ponders whether it is right to make money off a grieving person. Chiang closes the episode inconclusively—with Natalie's finger hovering, permanently in suspension, over the button that would transfer money to her account, leaving readers unable to determine what Natalie decides to do. We may at first infer that she does *not* press the button, for, in the subsequent episode in which she attends the Narcotics Anonymous meeting and speaks of doing a good deed, she may be referring to the transaction with Scott. However, the episode is followed directly by the one in which Dana receives the package containing videos of badly-behaved para-Vinessas and realizes that "[i]t must have cost a fortune" to compile them (Chiang 2019, 339). Which kind act did Natalie perform? Did she give the prism to Scott without taking the payment? Or did Natalie take Scott's money and use it to help Dana? Elegantly and playfully, Chiang emphasizes the many-worlds theory that is the impetus for the story, reminding us that Natalie's actions may be subject to quantum branching. In one branch, she would do a good deed for the grieving Scott, in another for the guilt-ridden Dana—but we cannot know. Chiang leaves us with an unopened box, Natalie's actions forever in superposition, unobserved and unmeasured.

References

Axvig, Max. 2019. "Quantum Mechanics, Contingency, and Freedom in Ted Chiang's 'Anxiety is the Dizziness of Freedom'." *Veritas Journal*, August 11. https://veritasjournal.org/2019/08/11/quantum-mechanics-contingency-and-freedom-in-ted-chiangs-anxiety-is-the-dizziness-of-freedom.

Chiang, Ted. (2002) 2010. *Stories of Your Life and Others*. New York: Vintage.

Chiang, Ted. 2019a. "Anxiety Is the Dizziness of Freedom." In *Exhalation*, 270–339. New York: Knopf.

Chiang, Ted. 2019b. *Exhalation*. New York: Knopf.

Chiang, Ted. 2019c. "Story Notes." In *Exhalation*, 341–50. New York: Knopf.

Clark, Tiana. 2023. "Maybe in Another Life." *The New Yorker*, September 11. https://www.newyorker.com/magazine/2023/09/18/maybe-in-another-life-tiana-clark-poem.

Counterpart. 2017–2019. Created by Justin Mark. Aired on Starz.

Faulds, Danna. 2002. "Foundation Stones." In *Go In and In: Poems from the Heart of Yoga*. Greenville, VA: Peaceable Kingdom Books.

Front, Sonia. 2015. *The Shapes of Time: Twenty-First Century Quantum Fiction*. Newcastle upon Tyne: Cambridge Scholars Publishing.

Gibson, William. 2014. *The Peripheral*.

Gibson, William. 2020. *Agency*.

Gleick, James. 1987. *Chaos: Making a New Science*. New York: Viking.

"Jackpot." 2022. *The Peripheral*. Season 1, episode 4. Directed by Alrick Riley. Story by Bronwyn Garrity. Teleplay by Scott B. Smith. Aired November 3, 2022, on Amazon Prime.

Lewis, David. 1986. *On the Plurality of Worlds*. Oxford: Blackwell.

McGinnis, Patrick. 2004. "Social Theory at HBS: McGinnis' Two FOS." *The Harbus*, 30 April. https://www.harbus.org/post/social-theory-at-hbs-mcginnis-two-fos-2.

McGinnis, Patrick. 2023. "How FOMO Became a Fixture." Interview. *Harvard Business School Alumni Stories*, May 11. https://www.alumni.hbs.edu/stories/Pages/story-bulletin.aspx?num=9169.

Miller, Andrew H. 2020. *On Not Being Someone Else: Tales of Our Undead Lives*. Cambridge: Harvard University Press.

Niven, Larry. (1989) 2001. "All the Myriad Ways." In *The Best Alternate History Stories of the 20th Century*, edited by Harry Turtledove and Martin H. Greenberg, 98–107. New York: Ballentine.

Parker, Jo Alyson. 2021. "Ted Chiang's Braided Stories of Memory, Narrative, and Technology: 'The Truth of Fact, the Truth of Fiction'." In *Performing Memories. Media, Creation, Anthropology, and Remembrance*, edited by Gabriele Biotti, 184–98. Cambridge: Cambridge Scholars Press.

Parker, Jo Alyson. 2022. "Ted Chiang's Time Travel Narratives: Predetermination, Predictability, and Free Will." *Science Fiction Studies* 49: 138–59.

Parker, Jo Alyson and Thomas Weissert. 2019. "Eternal Recursion, the Emergence of Metaconsciousness, and the Imperative for Closure." In *Time's Urgency*. The Study of Time 16, edited by Carlos Montemayor and Robert Daniel, 131–56. Leiden: Brill.

Phillips, Adam. 2012. *Missing Out: In Praise of the Unled Life*. Farrar, Straus and Giroux-Picador.

Rothman, Joshua. 2020. "What If You Could Do It All Over?" *The New Yorker*, December 14. https://www.newyorker.com/magazine/2020/12/21/what-if-you-could-do-it-all-over.

Sartre, Jean-Paul. 1956. *Being and Nothingness: An Essay on Phenomenological Ontology*. Translated and introduced by Hazel E. Barnes. New York: Philosophical Library.

Saunders, Simon, et al. (eds.). 2010. *Many Worlds: Everett, Quantum Theory, and Reality*. Oxford University Press. Online edition. https://doi.org/10.1093/acprof:oso/9780199560561.001.0001.

Schrödinger, Erwin. (1935) 1980. "The Present Situation in Quantum Mechanics: A Translation of Schrödinger's 'Cat Paradox' Paper." Translated by John D. Trimmer. *Proceedings of the American Philosophical Society* 124, no. 5 (October 10): 323. https://www.jstor.org/stable/986572.

Setinya, Kieran. 2017. *Midlife: A Philosophical Guide*. Princeton University Press: Princeton and Oxford.

Sterling, Bruce and Larry Shiner. (1985) 2001. "Mozart in Mirror Shades." In *The Best Alternate History Stories of the 20th Century*, edited by Harry Turtledove and Martin H. Greenberg, 296–312. New York: Ballentine.

Turtledove, Harry. 2001. Introduction. In *The Best Alternate History Stories of the 20th Century*, edited by Harry Turtledove and Martin H. Greenberg, vii–xiii. New York: Ballentine.

Turtledove, Harry and Martin H. Greenberg (eds.). 2001. *The Best Alternate History Stories of the 20th Century*. New York: Ballentine.

Wittenberg, David. 2013. *Time Travel: The Popular Philosophy of Narrative*. Fordham University Press.

CHAPTER 14

The Timekeeper and the Hour Glass: Artist Moving Image and Poetic Nonlinear Aspects of Time

Karen Heald

Abstract

In *The Timekeeper and the Hour Glass* the author, an interdisciplinary artist and academic whose practice embraces moving image explores poetic nonlinear aspects of time.

The research is set within the context of contemporary art, arts and science research, feminist action, and site-responsive approaches. Through a complex interdisciplinary methodology incorporating the performative use of time-based media while engaging with philosophical enquiry, visual analysis and experimental research, the exploration here draws upon Julia Kristeva's notion of intertextuality and Hito Steyerl's artworks and essay documentaries. The paper aims to articulate the phenomena of culture and the unmeasurability of temporal and spatiotemporal objects of sensory experience as distinguished from the clock.

Keywords

Interdisciplinary art practice – artist moving image – nonlinear narratives – experimental film – site-specific, visual analysis – in-between-ness – transitory strata – Julia Kristeva – Hito Steyerl – semiotic chora – spatiotemporal objects

1 Time and the Machine: Perceptions of the Clock

For many Western cultures the artificial construction of time is primarily measured through our perceptions of the clock. Clocks are a human construct designed to measure time, enabling us to retrieve information about the duration of events, and when and in what order they happen. Time therefore is significant within the context of the cosmos as well as in our own personal and social lives. However, attempting to decipher time's characteristics and

 | DOI:10.1163/9789004723634_016

properties, artistically, philosophically or scientifically has led to unresolved issues and yet more questions.

As an interdisciplinary artist and academic whose practice embraces moving image, I created a new short film *The Timekeeper and the Hour Glass*, described in this contextualizing essay, in order to explore poetic nonlinear aspects of time. My research is set within the context of contemporary art, arts and science research, feminist action, and site-responsive approaches. Through this complex interdisciplinary methodology I incorporate the performative use of time-based media while engaging with philosophical enquiry, visual analysis and experimental research. Drawing upon Julia Kristeva's notions of the semiotic 'chora' and 'intertextuality', as well as Hito Steyerl's essay documentaries, I use concepts of 'in-between-ness', 'being-in-between' and 'transitory strata' to inform my own arts practice, I position myself as an artist thinking-with Kristeva through the medium of film, seeing poetic language as the embodiment of language's materiality. To a lesser extent I also work with Steyerl's ideas of verticality and the poor image. I discuss here the work of these theorists and contemporary artists alongside my dream films and other work as a means of exploring poetic nonlinear aspects of time. In this way, as in my artwork, I aim to articulate the phenomena of culture and the unmeasurability of temporal and spatiotemporal objects of sensory experience as distinguished from the clock.

2 Research Questions

As part of the process of preparing for my return to Japan after almost two decades, I reflected upon my imminent visit in the context of the time passed. This influenced my choice of footage for the film and conference paper, while a sense of displacement that I experienced on arrival in Japan enabled me to further reflect on my practice-based research. The result was the following research questions:

- How has temporality evolved in my practice, from early experiments to my newest film?
- How does my artwork perturb commonsense versus scientific notions of time?
- How do the tripartite ideas of in-between-ness, being in-between, and transitory strata, contribute to a new filmic landscape?
- How is this configuration new?

I aim to explore these questions by situating *The Timekeeper and the Hour Glass* within my work overall and in relation to these theoreticians.

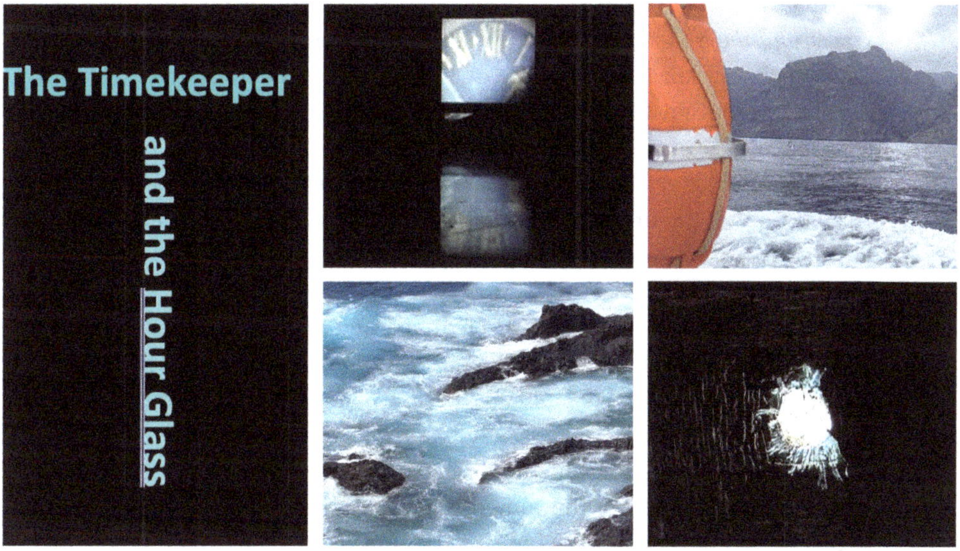

FIGURE 14.1 *The Timekeeper and the Hour Glass*, moving image duration 10′06″. Czech Republic, Tenerife, Wales

2.1 *The Timekeeper and the Hour Glass*

The Timekeeper and the Hour Glass is an artist moving image[1] which I created specifically for the International Society for the Study of Time (ISST) Summer 2023 triennial conference which took place in Yamaguchi, Japan on the theme of *Time and Measure*. The artwork was initially envisioned as an installation[2] involving multiple screens and film segments projected onto a variety of significant objects, alongside a contextualizing performative paper. As artist IIya Kabakov notes: “The main actor in the total installation, the main centre toward which everything is addressed, for which everything is intended, is the viewer.” However, due to equipment limitations, which affected all seven artists who participated in the exhibition and conference, the artwork was shown as a single channel film, on a timed loop with the other artists’ films. Follow the link below to view the film in its first iteration: *The Timekeeper and the Hour Glass* https://vimeo.com/841515905.

---

1 The genre of *Artist Moving Image* is situated between visual arts, and experimental film. For an expanded definition see “Defining Artists’ Moving Images” (2024).

2 The term installation art is used to describe large-scale, mixed-media constructions, often designed for a specific place or for a temporary period of time. For a more detailed discussion see “Art Term.”

Being restricted to the use of a single channel when exhibiting *The Timekeeper and the Hour Glass* enabled me to experiment with the flat 2D projection screen. The space in between, also known as the Ma in Japanese, refers to the gap or emptiness that exists between objects, situations, or relationships. It represents the essence of energy or intention that can be felt or experienced in that particular space, or a pause in the movement or flow. Physicists have discovered that the universe is made of energy waves and that every moment of our reality is a wave function collapsed from probability by a conscious observer. This means that our world is a product of our consciousness. Realities are not fixed but ever-changing creations we bring forth, both individually and collectively, through our beliefs and actions. As astrophysicist James Jeans put it: "The universe looks more and more like a great thought rather than a great machine" (1931, 137).[3]

By assimilating the blue hues and tonal qualities of the sea, *The Timekeeper and the Hour Glass* invites the audience to experience water as a spatial entity. In this way it also explores the in-between-ness I have been exploring throughout my artistic research, as the idea evolves into spacial form, particularly in terms of water, a fluid entity that evokes a place that is not a place. In contrast to Marc Augé's (1995/2023) concepts of non-places and supermodernity, spaces of transience where individuals remain anonymous and do not establish social bonds, this non-place or in-between-ness includes viewers by encouraging them to bring their own interpretation to the piece.

Water is estimated to be one of the most abundant molecules in the universe. Like most living systems it also displays a high degree of coherence. This has led researchers and philosophers to connect entanglement with numinous feelings of connectedness and conscious awareness in a way similar to Carl Jung's synchronicity and collective unconscious. In terms of memory and imagination, Claudiu Murgan's science fiction novel *Water Entanglement* (2018) asks when will we understand that water has memory? *Water Lore Practice, Place and Poetics* (2022) focuses on a number of affinities between water and imagination in contemporary fiction and nonfiction, asking what it would be like to attempt a thinking-with-water. All of these ideas make up a framework for my thinking-with Kristeva through film, in terms of the semiotic chora, intertextuality and my use of the tripartite ideas of in-between-ness, being-in-between and transitory strata.

---

3 In the *Mysterious Universe* Jeans (1931) was evidently aware of Plato's early philosophical perplexities as well as more recent ones of quantum theory, general relativity and quantum mechanics.

Exploring entanglement through a philosophical, aesthetic lens, Alva Noë, debates styles of seeing for painting and the pictorial arts. In trying to describe the phenomenon, he states: "Artworks—keeping with pictures—are puzzle objects in all the ways that the pictures in the newspaper or in the ad only exceptionally *ever* are. Indeed, they are, I would go *so far as to say,* philosophical objects" (2023, 43). Noë continues by describing pictures as proxies, a kind of graphene and dictograph comparative to written words and ways of using them: "We treat the mark on paper as if it were the word, and thus we use the device as a token of thinking about and better understanding ourselves as users of the word. This substitution is not exactly conscious, but it is not unwitting or misguided either" (2023, 58–59).

Working with this idea of juxtaposition, *The Timekeeper and the Hour Glass* embraces several sensitive forms depicted in unexpected scenarios, such as the slow unfolding of an underwater fern, projected upwards towards the aperture of the circular domed roof of a medieval dovecot, creating the illusion of the antithesis of an abyss. The warm turquoise and deep blues of the seascapes, and the subtle grey hues of the distant hills and mountain peaks, contrast vividly with the blacks and dark browns of interior architectural spaces and contemporary galleries. A quote from Tacita Dean, appropriated and transcribed over these images, establishes and disrupts the horizon through the text's fragmented appearances.[4] The experimental sound created by Susan Matthews for the film establishes a discord with the images, producing a tension that emphasizes the cool blue of the film's intended color palette.

I am intuitively drawn to the color blue within my films, responding to Kristeva's account of the color as well as Susanne Ferwerda's, in her article, *Blue Humanities and the Color of Colonialism* (2024). Ferwerda analyzes the lapis lazuli series by Dutch artist Pieter Paul Pothoven and the performance of the poem *Unity* by Aotearoa New Zealand poet Selina Tusitala Marsh, to argue that blues address colonial inheritances and critique colonial desires. As Pothoven's work shows, the color blue has built empires, taken lives, and altered environments. In contrast, Kristeva writes about the colors used in Giotto's frescoes, and in particular how the overwhelming shades of blues translate "instinctual drives into colored surface" (1980, 210). Kristeva draws on color perception theory and infant vision development to argue that blue has a quality beyond identity. A possible hypothesis, following Andre Broca's paradox that (to see a blue light, you must not look directly at it) "would be that the perception of blue entails not identifying the object; that blue is, precisely, on

4 See Yentob (2018, 24:37).

this side of or beyond the object's fixed form; that it is the zone where phenomenal identity vanishes" (qtd in Ferwerda 2024, 7).

The role of location within my films has also always been significant, particularly place and site-specificity. In creating *The Timekeeper and the Hour Glass*, the locations, which included the Czech Republic, Tenerife and Wales, synthesized my original analogue films shot over 20 years with current digital footage. This was the first time I had combined different film media footage to create a new non-linear narrative, and one that reflected, both spatially and temporally, the juxtapositions of my own life.

My filmmaking is a process of writing as much as recording, echoing the automatic writings of André Breton and the surrealist circle, particularly the women surrealists. There is a decisive act of picking up the pen, switching on the computer, pressing the shutter or record button, and then a release into an imaginary world. I therefore relate to the intuitiveness of working in film and artist moving image. The process prompts thinking through the lens. My practice occurs throughout the duration, from the scoping of initial ideas, locations and people to synergizing intuitive and reflective practice in the editing, presentation and dissemination processes.

This conditional framework has enabled my practice to remain intuitive, reflective, and also responsive while operating upon different levels of consciousness. Through the unique use of the video camera, the editing process, and analogies and cross-readings between theory and practice, my work, beginning with my creative experimental dream films, explores non-linear preverbal spaces. These films visualize a re-interpretation of sleep, the unconscious and experiential time and spatial elements through audio-visual technologies. *The Timekeeper and the Hour Glass* drew upon and developed out of this work which began with *Drift*, staged sleep performances created in the Sleep Research Unit at the University of Charleroi, Belgium, with Dr Myriam Kerkhofs in 2004. They then continued with a Wales Art International (WAI) funded project, *The Dream Space*, during my first visit to Japan in 2007.

2.2 *The Dream Space*

While on my research trip to China, Mongolia and Japan in 2007, I travelled via the Trans-Mongolian railway. My journey through Asia gave me the opportunity to experience time and sleep in different locations and to explore its meaning to other cultures. This relationship to travelling became paramount in my photography and film work. By investigating the politics of location, I also examined what sleep and the body meant cross-culturally and in different spaces.

FIGURE 14.2 *The Dream Space: Drifting on a Pink Cloud*, moving image, duration 9′28″. 2007. Japan

FIGURE 14.3 *The Dream Space: Yoro Park*, performance installation. 2007. Japan

FIGURE 14.4 *The Dream Space: MOBIUM*, mobile museum. 2007. Japan

Through lens-based media, I explored time and what it might mean as a woman to be staging sleep performances in different spaces and locations. Sleep was looked at from a physical viewpoint, and through the unconscious and imagination through dreaming. As my interest in dreaming evolved, coming to fruition in Japan through research for *The Dream Space*, I became increasingly interested as well in how dreaming allows entrances to simultaneous time zones.

The photography and films created as part of *The Dream Space* explored time along with sleep experiences in contrasting spaces in Japan, from the capsule hotels designed for speed and mass capitalization, to a Buddhist temple, requiring reflection, ritual and spirituality. The work also questioned the notion of a separation between observer and observed, as I investigated the varying viewpoints of the camera lens and explored what happens when women are on both sides of the camera.

The Dream Space was a collaborative project with visual and sound artists Rachael Kearney, Takeko Akamatsu and Yosuke Kawamura. The preliminary artworks were exhibited and toured Japan via MOBIUM (a mobile museum). Venues included the Museum of Modern Art, Kyoto and the University of Nagoya.

3 Tripartite Structure: In-between-ness, Being In-between and Transitory Strata

As part of the creative research process, the dream films explored and visualized Kristeva's concepts of semiotic chora and monumental time. Reflecting upon these ideas within the digital revolution then gave rise to the concepts that I see as forming a tripartite structure and identify as in-between-ness, being in-between and transitory strata. In-between-ness refers to a conceptual, reflective public space within collaborative arts and science practice. In this sense, in-between-ness is a system which allows an interdisciplinary and multidisciplinary methodology through whose approaches and media two different subjects may meet. As such the practice moves between different modes, responding uniquely to the qualities of individual artists and institutions.

Being in-between is a space that oscillates between the public places of in-between-ness and the personal preverbal space of transitory strata. Transitory strata are preverbal, personal spaces which draw from Kristeva's semiotic chora and which, through the making of experimental dream films, visualize a feminist re-interpretation of passive/active sleep, the unconscious, and experiential time. This differs from the split or divided self as it pertains to fleeting parallels of time that exist within the creatives or the depressives state of being. This ephemerality is discovered within the individual, lying amidst the layers of consciousness, the realms of waking, reverie and wakefulness. These experiences seldom occur whilst asleep, but can be stirred by direct sensitivity to sounds, which, when heightened under extreme moments of creativity, stress or illness, can take a person on a more profound pathway through abstract associations and internal ruminations. These journeys, or transitory strata, have the ability to both build and fragment, and the dream films aim to echo this on the level of the unarticulated and the unutterable.

Returning to Japan 17 years on from my initial artist's residency brought with it a feeling of displacement, not only from the extremely visceral effects of my inert circadian rhythm and that of others surrounding me, but also from the sense of being in-between. Creating new footage and films whilst in Japan in 2023, and ruminating on the unmeasurability of time, I was also, in solitary moments, further advancing the concept of transitory strata. The results are to be exhibited in various iterations on time in collaboration with sound artist Susan Mathews, from May 2025.

In the following sections I consider the work of some visual artists and cultural theorists through the tripartite structure of in-between-ness, being-in-between, and transitory strata. My aim is to explore the unmeasurability of time through poetic nonlinear methodologies. For example, in terms of the visual artists, my idea of the split or divided self recalls Trinh T. Minh-ha's practices, while the idea of layers of consciousness speaks to Tacita Dean, both as supporting my temporal configuration, and as revealing where my work moves beyond their conceptions of time.

3.1 *Context*

I have established various approaches to collaborating with people and sites, mediating between the physical act of filmmaking, the subconscious realm of knowledge, and arts and science research. While filming in unfamiliar locations or conditions, the process commences with an initial plan but always allows for serendipitous situations to occur.

This practice charts a multifaceted research journey. The initial emphasis focused on cyclical, maternal, and monumental time; to synthesize scientific debates in sleep, the unconscious, and melancholia with current dialogues in women's films. In order to investigate the spaces that lay between these states I engaged with Julia Kristeva's key texts, *Revolution in Poetic Language* (1974) and *Women's Time* (1979), examining women's sleep from a feminist, artistic and scientific gaze, both through an initiated Artist Residency, *Drift*, at the Sleep Research Centre of the Charleroi University Hospital and then in the film *Drifting on a Pink Cloud*, which formed part of *The Dream Space*, touring exhibition in Japan.

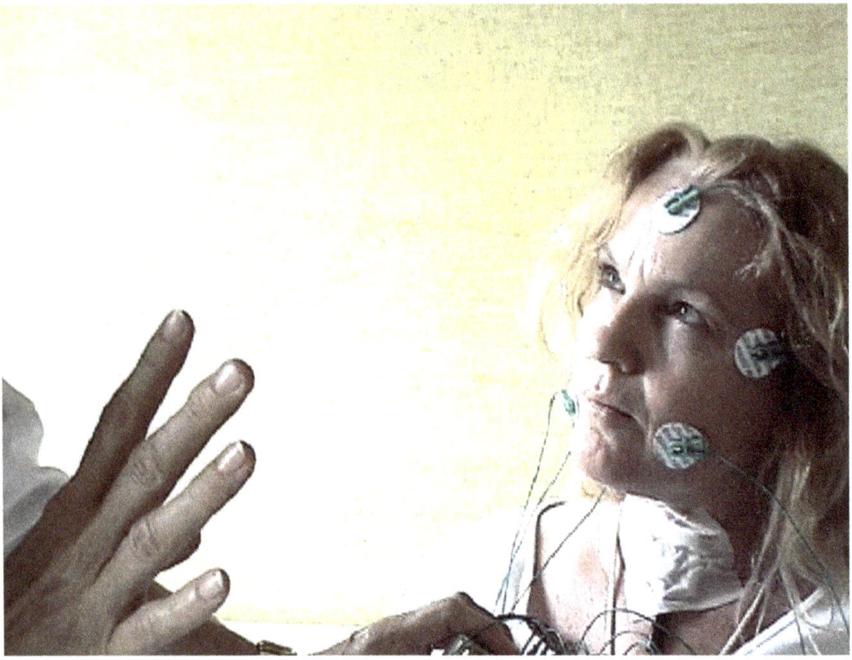

FIGURE 14.5 *Drift*, Artist Residency, Sleep Research Centre, Charleroi University Hospital, Belgium

By studying the poetics of image and text in the in-between states of wake and sleep, women's time was visualized via feminist cultural theory on time and location.

In exploring Kristeva's semiotic chora, women's time and intertextuality my dream films are created in cycles and staged in a variety of locations and in-between spaces such as on the Trans-Mongolian Railway, in a capsule hotel in Japan, and in a council flat in East London. The intention is to offer the viewer a translation of how we may encounter and be affected by time, just as time, in all its varieties, appears as a mechanism through which to formulate the in-between-ness of significant artistic methodologies such as collaborative practice.

The methodological approach of collaborative arts and science practice initially drew on time and Kristeva's idea of the chora as a preverbal and creative space relating to rhythms, colors and trace. This initial stage takes the form of filmed performances which explore the physicality of the female body within site-specific spaces and staged performances. I was specifically fascinated with the fragmentation and non-linear aspects of time within dreams, the stillness of the body and activity of the mind, and the unmeasureabilty of time in this context, given that most dreams are only recalled after the event, upon waking. These issues are also central to Kristeva's interest in how people operate in the symbolic order.

While Kristeva, as a linguist and psychoanalyst is interested in how to deal with the line between the symbolic and the semiotic, using music to help her patients in clinical settings, my methodological approach is not intended as art therapy. Rather I use contemporary art, specifically video and artist moving image, as disciplines in their own right and as creative research. Preverbal and multiple intelligences, such as visual, kinaesthetic and tactile learning are crucial to the study and introduced through objects, which were inscribed with prompts.

My interest in the in-between continued in my work with patients with mental health problems as part of my Artist Residency *Visualising the Invisible* (2009–2010) with Arts Council Wales and the Welsh National Health Service (NHS). Here my research moved from the observation of the physical body to explorations into the invisibility of the mind, using video as a creative tool. This subsequently led to an interdisciplinary three-year art and science funded research project called *In-between-ness* with Arts Council Wales, North Wales Clinical School and NHS Wales, collaborating with artists, psychiatrists and people living with major depression. Paradoxically, it is through the abstract non-linear narrative of dream films that the patients begin to contextualize their world.

FIGURE 14.6 *In-between-ness*, visual prompts. 2010–2013

The in-between-ness study entailed a three-way process developed between the artists/participants, participants/psychiatrists and psychiatrists/artists. Attempting to create a non-hierarchical relationship in this situation required collaboration and co-creation, which enabled the participants to have authorship of their own artworks. In this way we sought to create an ethical and philosophical awareness of the power-relations inherent in different roles in a hospital, here, for example, those of participants and psychiatrists.

My next work, the artist moving image *Paper Interior* (Karen Heald and Susan Liggett 2013), was created in response to the three year *In-between-ness* arts and science research project. Filmed in daylight in a former hospital—currently a Fine Art institution—*Paper Interior* was performed live as part of the evening preview event of the exhibition *In-between-ness: Using Art to Capture a Sense of Self*. It was inspired by one of the prompts used in the project, to document your day in a variety of ways, and referenced Guy Sherwin's *Paper Landscape* (1975). The film recalled a reverie in its lighting and interior/exterior framing,

FIGURE 14.7 *Paper Interior*, moving image, performance, installation 2013. Oriel Sycharth, Wales

designed to create a physical, metaphorical and philosophical in-between-ness of space.

Having been first performed live to an audience, the film further incorporated the temporal and spatial as it was later viewed as a staged performance where the film screen was virtually and literally painted, slicing through conscious and unconscious time. *Paper Interior* thus explored different concepts of time and perceptions of reality, in this case for those undergoing antidepressant treatment, a situation where identity is not fixed but slips between different realms of reality. We aimed to evoke a sense of in-between-ness in an attempt to visualize the ephemeral and transitional spaces of the mind. As we literally sliced through the screen and stepped into the gallery space there were gasps from the audience who had not anticipated this action and its configuration of the tripartite in-between-ness, being in-between, and transitory strata.[5]

The dream films are therefore key to the overall process of my work and its paradoxical undertaking of giving a voice to the unutterable, or, perhaps rather, allowing the unutterable to disperse more widely and with a sense of authority and agency. In this way it responds to Gayatri Spivak's work in *Can the Subaltern Speak*? (1988), which critiques the tendency of Western intellectuals to speak on behalf of subaltern groups (those who are socially, politically, and economically marginalized), often without allowing them to represent themselves. Spivak's work intersects with the visual in dialogue with artwork by Ecuadorian artist Estefanía Peñafiel Loaiza (2021), whose practice questions visibility and occlusion, and the need for and absence of the image.

Like Spivak and Loaiza, Kristeva recognizes the power of art to unblock important sources of meaning. Although philosophy and psychoanalysis often privilege language and conceptual distinctions, mistrusting the image, Kristeva, a practitioner of both disciplines, explores the ways in which creative acts may counteract and transform feelings of violence and depression on both the individual and the cultural levels. Similarly, my performative use of time-based media, established through a complex collaborative methodology continues to move forward within the arts and science arena, bringing together psychiatric issues, collaboration and film.

My research with creating moving image further expands on Kristeva's chora in relation to time and the concepts referred to earlier: in-between-ness, being in between and transitory strata. These concepts relate to the different spaces within which I operate, such as the space between arts and science disciplines, the collaborative space of working with different people, and the space

5 See Heald (2013). View *Paper Interior* in its entire duration of 9′20″ by following the link below: Paper Interior Vimeo: https://vimeo.com/74064501.

between sleep and dreams. Each environment commences with dialogues with people or places, establishing the parameters of the following stages as various iterations develop the uniqueness of the diverse dream films. My artist philosophical investigation of time thereby investigates the complexities between commonsense views of time and what is viewed through the lens of arts and science, particularly visual arts and physics.

While ideas of time are generally expressed in "non-technical terms such as *now*, *flow*, and *past* and not with technical scientific terms such as *continuum*, *reference frame*, and *quantum entanglement*" (Dowden 2023) we are aware that there may be many further dimensions beyond our ordinary experience.

Quantum entanglement in physics is a phenomenon where the properties of two particles are correlated so that measuring one particle instantly affects the state of the other, regardless of the distance separating them. This idea defies classical notions of locality and has led to intriguing experiments and philosophical debates.

Within my practice something like quantum entanglement occurs through the complexity of film, a medium that operates within time while also disrupting time through its contextual referencing, capture (framing), and use and suspension of duration, pause and tempo through mirror or mirrorless mechanisms. When film enters the realm of installation and artist moving image the dynamics shift even further, to include the contemplations of the audience, readers, viewers.

Like scientific objects, such as the camera and/or our ubiquitous smartphones, our human eyes are internally operated by our brain, entangled within their own time and space. When I am filming, I am fully engaged; I think through the lens. There is something about the time and space between the eye and the lens of the camera that enables me to do so, for example, when working with dancers, the entwinement of bodies and the connection that exists between us as humans, or as when interacting with non-humans or attempting to become non-human, we are entwined together in spontaneous and intuitive interactions that exist as experiences within our memories. As an artist-researcher my focus is on visual, tactile and auditory sensations within our perceptions of reality. As with dream spaces which exist through non-linear narratives focusing on different elements—color, rhythm or movement—these operate within different multidisciplinary and interdisciplinary languages which are forever entangled. Hence my continuous paradoxical experiments and explorations within and without space through time-based media[6] aiming to visualize the

6 Time-based media is a term used by museum curators to describe art that has a duration or unfolds over a period of time. It can include video, film, audio, light, movement, and performance. Time-based media can also be called new media.

invisible. Returning to Noë's comments on entanglement and debates on styles of seeing for painting and the pictorial arts, it seems appropriate here to see artworks—as pictures—as puzzle objects. The connections between serendipity, locations, and site-specific artworks are thus timely given our fluid and ever changing conceptual, ecological, social, and cultural environments.

3.2 *Further Dream Films*

In 2013, following the In-between-ness pilot study, I was invited by interdisciplinary artist Angela Davies to collaborate on an artist residency called *Golau* (*Light*) in Chirk Castle and its gardens and grounds, located in north Wales. Here I explored themes of time and place in response to the architectural framework and geologically sculpted landscape of Chirk Castle, a 12th Century medieval fortress in the Welsh Marches, set within 480 acres and under the auspices of the National Trust. As part of the residency, I collaborated with Angelina Kornecka, an artist and trained Butoh dancer, to create 5 site specific film performances over two intense days. In *Topiary* (Heald, Davies, Kornecka 2014), Kornecka appears grounded, walking along the top of the tall hedgerow in the estate gardens of Chirk Castle. Her movements are slow and deliberate yet precarious, both for her and for me as I filmed the process. The horizon is unstable as Kornecka navigates and traverses the site, meeting herself through cinematic montage, mirrored in screen time. The film offers a diptych of chronological and poetic non-linear time as the virtual present attempts to meet the self, obstructed by the phallic shaped cut topiary.

FIGURE 14.8 *Topiary*, filmed performance. 2014. Chirk Castle, Wales

FIGURE 14.9 *Lateral Flight*, filmed performance, duration 3′22″. 2014, Wales

Golau (*Light*) was designed to generate multi-sensory and multidisciplinary responses to the memory of place, here focalized through Chirk Castle. Built in 1310 during the reign of the conquering Edward I, the castle was a symbol of power and a statement of English intent in these disputed lands. The landscape sits between the rivers Dee and Ceiriog along a preserved section of Offa's Dyke, a large linear earthwork that roughly follows the border between England and Wales. Within the wooded grounds lies an old folly which contributed to my site-specific installations and performative works, *Lateral Flight*, *Golau* (*Light*), and *Meander*.

The process of the collaboration was an inspiration. Filming in the interior and exterior spaces of Chirk Castle, I worked intuitively with Kornecka. *Lateral Flight*, a film performance that experimented with the horizontal and vertical planes of the film screen, used emotive movements to capture shifts in light and portray intimate moments in time through the camera lens. The editing and presentation process reiterated the Butoh sense of meditation, minimalism and being in-between.

To return to Steyerl:

> Imagine that you are falling but there is no ground. Many contemporary philosophers have pointed out that the present moment is distinguished by a prevailing condition of groundlessness. We cannot assume

> any stable ground on which to base metaphysical claims or foundational political myths. At best we are faced with temporary contingent, and partial attempts at grounding. But if there is no stable ground available for our social lives and philosophical aspirations, the consequence must be permanent, or at least an intermittent state of free fall for subjects and objects alike. But why don't we notice?
>
> 2012, 13

Reflecting this intermittent state of free fall for subjects and objects alike, I filmed Kornecka in *Lateral Flight* in a continuous long take, as her body slowly and vertically slid down the twisting medieval stairs of the castle, with the silk fabric intermittently rustling against the worn stone steps. In the installation, the film *Lateral Flight* was projected onto a circular photography reflector, reframing the film's form and interaction with the castle's interior architecture and the exterior diffused light entering from the window frame, aiming at a visual re-interpretation of site, the unconscious and experiential time.

FIGURE 14.10 *Lateral Flight*, moving image, installation. 2014. Chirk Castle, Wales

FIGURE 14.11 *Golau* (*Light*), moving image, installation. 2014. Chirk Castle, Wales

FIGURE 14.12 *Meander*, filmed performance, installation. 2014. Chirk Castle, Wales

The themes of *Lateral Flight* were reiterated in *Golau* (*Light*) this time using open-weaved muslin, selected for its opacity, to capture the earlier filmed performance of Kornecka's figure slowly traversing the folly in the grounds of the castle. The installation, like the performance, was fleeting and ephemeral, depending upon the light and changing weather conditions during particular times of, and on particular days. Here, the light from the golden hour pierced through the window and fabric, deflecting onto the walls and floor. The intention here, as in the other dream films, is that the viewer experience a translation of how we may encounter and be affected by time in the in-betweenness of filmmaking and research as collaborative practice.

3.3 *Dream Films, Time, Steyerl, and Kristeva's Writings*

Hito Steyerl, a German-Japanese filmmaker, moving-image artist, writer, and innovator of the essay documentary is a prolific conceptual artist, a professor of experimental film and video, and the co-founder of the Research Centre for Proxy Politics at the Berlin University of the Arts. As noted above, the main concepts that I have been drawn to are her theories of verticality and of the poor image. Steyerl (2011) identifies the present moment as a state of free fall, distinguished by a prevailing condition of groundlessness:

> While falling, people may sense themselves as being things, while things may sense that they are people. Traditional modes of seeing and feeling are shattered. Any sense of balance is disrupted. Perspectives are twisted and multiplied. New types of visuality arise. This disorientation is partly due to the loss of a stable horizon. And with the loss of horizon also comes the departure of a stable paradigm of orientation, which has situated concepts of subject and object, of time and space, throughout

> modernity. In falling, the lines of the horizon shatter, twirl around, and superimpose.

This loss of stability connects also to Steyerl's idea of the poor image which occurs between digital and analogue processes through the degradation of mass-reproduction and distribution through video as a medium, with reference as well to smartphones. The idea also raises questions about imperfect cinema, visual bonds, and their relationships to poor images and society:

> The poor image is no longer about the real thing—the ordinary original. Instead, it is about its own real conditions of existence: about swarm circulation, digital dispersion, fractured and flexible temporalities. It is about defiance and appropriation just as it is about conformism and exploitation. In short it is about reality.
>
> STEYERL 2012, 44

As a filmmaker, I relate the ideas of free fall and of the poor image to the time and space of being in-between, a feeling of being in two worlds simultaneously, a liminal space between reality and consciousness. Time is viewed here as an interface between two dynamic states, somehow just beyond the visible. My exploration of time and the perceptions of the clock in *The Timekeeper and the Hour Glass* also reflects Julia Kristeva's multilayered sense of time, such as cursive time (the time of linear history) and monumental time. As Ananya Tiwari (2017) puts it, for Kristeva: "Monumental time is not a time but rather a particular space where a particular mindset permeates, and can transcend time, space, socio-political and economic conditions, nations etc., which diagonally to each other across [sic] time and space. Symbolic denomination is a part of monumental time."

As Kristeva writes, "the fact that two types of temporality (cyclical and monumental) are traditionally linked to the female subjectivity in so far as the latter is thought of as a necessarily maternal should not make us forget that this repetition and this eternity are found to be fundamental, if not the sole, conceptions of time in numerous civilizations and experiences, particularly mystical ones" (1979, 17). Thus women's conditions can be linked across the spectrum, via monumental time, irrespective of linear time and socio-economic/national reality.

Kristeva's concept of the chora as semiotic, that is, as the space in which meaning is produced, has acted as a model for my research. Kristeva uses maternal language when discussing the chora, comparing it to the contractions of giving birth and the birth itself. She describes the "'energy' charges as well as 'psychical' marks, [which] articulate what we call a chora: a non-expressive totality formed by the drives and their stases in a motility that is as full of movement as it is regulated and cannot be represented" (1984, 25). Examples might include the preverbal sounds of an infant, the psychotic who has lost the ability to use language in a meaningful way, or the artistic avant-garde, which mobilizes the semiotic in its push for originality, creativity and critique.

The complexity of Kristeva's theories in relation to linguistics and feminist critical theory has challenged me to create the unrepresentable or the unspeakable through moving image. In collaborations with artists operating in a variety of disciplines, I have critiqued and engaged with Kristeva's propositions to give a re-interpretation of time.

Tacita Dean is renowned for such pioneering and poetic 16mm and 35mm films as *FILM* (2011) and *Antigone* (2018), which experiment with the medium of celluloid film and narrative. Dean frequently employs rotational or circular movements within her films, which act as indicators of time in an actual, cinematic and psychological, coming together through unconventional framings of familiar scenes. In earlier films such as *Disappearance at Sea* (1996) Dean captured close-ups of the rotating lighthouse bulbs while the distant horizon is only glimpsed through the glass. In *Fernsehturm* (2001) Dean filmed a restaurant inside the rotating TV tower in Berlin using 16mm film and an anamorphic lens which achieved a wide angle and long rectangular view. Filmed for 44 minutes at twilight it captured both the changing exterior light and the interior narratives. Although reminiscent of *Disappearance at Sea* in its focus, here the framing resides on the restaurant's interiors, with the primary vista seen only from a sideways perspective, punctuated with gestures from diners pointing outwards. Describing the work, Dean wrote:

> Like the perpetual rotation of the spacecraft in Stanley Kubrick's '2001: A Space Odyssey', a conceit to maintain gravity on board ship, the *Fernsehturm* restaurant continues to turn almost imperceptibly like the movement of the planets in Space. It was visionary in its concept and a symbol of the future, and yet it is out of date. The *Fernsehturm* embodies the perfect anachronism … As you sit up there at your table, opposite the person whom you are with, and with your back to the turn of the restaurant, you are no longer static in the present but moving with the rotation of the Earth backwards into the future.
>
> DEAN AND GROENBOOM 2000

Dean's practice encompasses an intertextuality of image and text informative for my idea of in-betweenness, not only through her film montages but also in her drawings, paintings, photographs, and installations. Her recent work,

FIGURE 14.13 Tacita Dean (2001), *Fernsehturm*, 16mm film, TV Tower, Berlin

Geography / Biography (2023), commissioned by the Bourse de Commerce in Paris, is a sophisticated collage of her life in film shown via two synthesized 35mm projectors mounted on a spinning platform with backgrounds from a personal collection of postcards depicting plant life, landscapes and ancient statues. A catalogue accompanying the exhibition acts as an additional archive of the work. Time here is her material[7] since alongside film montage of Dean's life, are interspersed the slow paced artist's film-portraits of elderly men such as Claes Oldenburg (2011) and David Hockney (*Portraits*, 2016), all imbued with a sense of waiting.

Just as Dean's work is informed by time, Minh-ha's book *D-Passage* (2013) documents her feature film *Night Passage* (2004) and installation *L'Autre marche* (*The Other Walk*) (2006). Reflecting on these works she discusses the impact of new technology on cinema culture and creative practice while exploring several of her major themes, such as temporality, transitions and transcultural encounters. The sensitivity with which she discusses ways of

7 Sean Burns. "Time is Tacita Dean's material." *Freize*, issue 237. EU Reviews, Exhibition Reviews. June 14, 2023. https://www.frieze.com/contributor/sean-burns.

FIGURE 14.14 Tacita Dean (2023), *Geography / Biography*, exhibition, Bourse de Commerce, France

seeing and knowing, along with the implications of the media used, the artistic practices engaged in, and the representations created, comes from the synthesis between theory and practice. *D-Passage* indivisibly weaves form and structure, rhythm and movement, language and imagery through essays, artistic statements, and in-depth conversations, all joined with the script of *Night Passage*, movie stills, photos, and sketches. This multi-disciplinary approach has also been very important for my own work.

4 Conclusion

Thinking through the research questions that I raised at the beginning of the chapter brings me also to work I was invited to create for *Heterotopic Encounters*, an exhibition at an artist studio, UnDegUn. The 2019 exhibition included colleagues and artists (former students) and invited audience engagement. The dialogue started with a quote from Michael Phillipson written in 1995:

> Under the intensity of creative experimentation (the situation and challenge of artists' practice) each work asks itself (and therefore us, too)

FIGURE 14.15 *Heterotopic Encounters*, moving image, installation. 2019. UnDegUn, Wales

> whether there might be a 'place' where culture has not yet reached; it hopes to be that 'place'—an elsewhere that is not yet a 'place' on culture's terms.[8]

I was, accordingly, interested in the visual dialogues and conversations between artists who reside within the artistic space of UnDegUn, both those collaborating and visitors experiencing the event. Given the recent precarious nature of the location, its temporary lifespan, and the early closure of the exhibition due to the COVID-19 pandemic that transversed the globe, this was especially poignant and significant.

The two films I exhibited had been brought together and re-framed for the first time at UnDegUn, in my hometown of Wrexham. The films shown explored the apertures and mechanics of the camera, the unconscious and the active/passive role of the female in the landscape.

The first film *Untitled IV* (2001) is a silent video, filmed along the Irwell Valley Sculpture Trail in Salford. It explores Salford's industrial heritage, the graphics of line, tone and form, Rosika Parker's *The Subversive Stitch*, and the idea that when sleeping the brain is still active. *Pra I* (2004) is one of two films that reflect upon early experimental Czech cinema; it was filmed through a viewing device along the river Vltava that flows through Prague, whilst attending an international conference on sleep. Both films are influenced by Julia Kristeva's concept of the semiotic chora and my notion of transitory strata and in being brought together many years later also incorporate time as such into the installation.

Installing the work I experimented and played with the projection screens, the bleeding out of light and sound, the piano score, the grain and aesthetic

8 Phillipson, Michael. 1995. "Managing Tradition: The Plight of Aesthetic Practices and Their Analysis in a Technoscientific Culture." In *Visual Culture*, edited by C. Jencks, 202–3. London and New York: Routledge.

tonalities of analogue films, and the objects themselves, sourced from artists and found in the space, all of which are to be celebrated. The interactive overlaps, such as the found piano which I brought into my space, acted as invitations for people to continue to play and experiment with the work throughout and beyond the exhibition.

Thinking about play and experimentation on returning to Japan and experiencing the people, landscape and culture of the country for a second time, brought new and distant memories, of loss, displacement, sensory experiences and new friendships. In the film *Forgetting Vietnam* (2016), Minh-ha, following her father's passing, explored the relationship between the human and nonhuman through the abstraction of love, which she described as encompassing more than feelings and emotions, and as that which connects us. In this film Minh-ha explores Vietnam, approaching: "… Her as a body of water—a reality manifested in the overall texture of Vietnamese life; a reality implied in the Vietnamese term nu,o,c, meaning both water and country".

Water has been a current theme within my work from as far back as 2001 when I created short films such as *Frozen Seas* and *Intangible Fluidity*, up to 2023 with this first iteration of the artist moving image *The Timekeeper and the Hour Glass*. For this new film I sourced my own film archive spanning over two decades and created a new film to synthesize analogue and digital arts footage from various locations including the Czech Republic, Tenerife and Wales. I experimented with expanding and compressing perspective and included

FIGURE 14.16 *The Timekeeper and the Hour Glass*, moving image, 10′06″. 2023. Wales, Tenerife, Czech Republic

FIGURE 14.17 *The Timekeeper and the Hour Glass*, film still. 2023. Wales, Tenerife, Czech Republic

text to subvert the flat two-dimensional screen. The disharmonious piano sounds were intended to further disturb and distort the visual and textual imagery presented to the audience in its multiple contexts on a single channel video—from the art exhibition to the video essay relayed at the conference.

My films frequently speak as well to a painterly textural aesthetic. These reflections on film, texture, smell and tactile knowledge appeared in *The Timekeeper and the Hour Glass* as I conjured up what a space in Japan with an international audience of interdisciplinary scholars and artists focusing on the theme of time and measure might be. One of my initial prompts was thus this quote by Tacita Dean (2000):

> One hears or looks at a film; but one can only literally touch, taste and smell a piece of celluloid. Perhaps it is necessary to return here to the notion of film as event rather than a mere spectacle, instead of centering on the screen, the viewer's experience of film is also engaged in the extra-screen space—that is, the movie-house space of the immediate environment.

I am reminded here of a practice in Japanese Kabuki theatre, devised to heighten the audience's sensual experience through the play with the extended notion of the stage and the many uses of passageways(s) (hanamachi).

FIGURE 14.18 *The Timekeeper and the Hour Glass*, film stills. 2023. Wales, Tenerife, Czech Republic

In *The Timekeeper and the Hour Glass* I fragmented Dean's quote into segments and superimposed the text onto the moving image to reimagine alternative spaces, particularly in unfamiliar locations. The rhythmic movement of the sea and at times its circular motions added to the drama and the serendipity of the occasion, enabling time and space to unfold so things or events can play out.

Minh-ha, in an interview with Shivani Radhakrishnan (2022), stated that despite feeling strongly on certain issues she doesn't enter projects with

FIGURE 14.19 *The Timekeeper and the Hour Glass*, film still. 2023. Wales, Tenerife, Czech Republic

preconceived ideas. However, a constant within her work is: "the question of cinema and of writing—how the creative tools at work define my activity in relation to a specific subject or a specific course of events. Or better yet, the way a work breathes, moves, pauses, and rests."

Interviewed by Alan Yentob, as part of the *Imagine* series, Dean ruminates on why we become who we become or do as we do, saying it remains for the most part unknown to us, the most mysterious question being why artists become artists, and why they become the artists they are. Like Dean, I do not want to know what the outcome is before I have taken the journey, and along with Dean I think that that is the one thing I can identify with in terms of our process. As Dean says, this makes the process harder but: "I think I need to have a bit of invisibility about myself …" (Yentob 2018).

The Timekeeper and the Hour Glass is an ongoing project in collaboration with sound artist Susan Matthews which will transform into a triptych installation and site-specific intervention containing audio/visual technologies and a variety of objects. This is intended for international contemporary galleries as well as relevant exterior locations in Wales and beyond. The re-contextualization and re-siting of the work is a significant part of the process which is designed to span several years and repetitions.

Time is a material which appears to become more unmeasurable and messy the further I explore it, creating more questions than answers. For example, in-between-ness is becoming spatial, a factor that is particularly powerful in

FIGURE 14.20 *The Timekeeper and the Hour Glass*. Panorama photograph. 2023. Potential installation site. Japan

The Timekeeper and the Hour Glass, which features unfamiliar locations. The spatial in-between-ness within my film conveys water as a kind of place that is not place, in contrast to Minh-ha's concept of Vietnam as a body of water or Dean's moving backwards into the future via the Earth's rotation. Through visual analysis of artist moving image and poetic nonlinear aspects of time this transition between the tripartite system, in-between-ness, being in-between, and transitory strata, can, with audience interaction, contribute to a new filmic landscape.

Could then the location above, visited by me while I was in Japan in 2023 become a potential new site for further iterations of *The Timekeeper and the Hour Glass* film?

The essay concludes with the weblink to the first iteration of *The Timekeeper and the Hour Glass* (2023) for viewing and processing by the reader/audience with comments welcomed. Follow the link below to view the film in its current single channel format: https://www.karenheald.co.uk/films.

References

"Art Term." N.d. Tate. Accessed October 27, 2024. https://www.tate.org.uk/art/art-terms/i/installation-art.

Augé, Marc. (1995) 2023. *Non-Places: An Introduction to Supermodernity*. Translated by John Howe. New York: Verso Books.

Bourdier, Jean-Paul and Trinh T. Minh-ha. 2006–2009. *L'Autre marche* (*The Other Walk*), installation. Musée du Quai Branly, France.

Burns, Sean. 2023. "Time is Tacita Dean's material." *Freize* 237, June 14: https://www.frieze.com/article/tacita-geography-biography-2023-review.

Dean, Tacita and Ronald Groenboom. 2000. *Tacita Dean*. Barcelona: MACBA.

Dean, Tacita. 2011. *FILM*. Unilever Series, Turbine Hall, Tate Modern, UK.

Dean, Tacita. 2018. *Antigone*. Royal Academy, London, UK + Touring.

"Defining Artists' Moving Images." 2024. Irish Artists Film Index. Accessed October 25, 2024. https://irishartistsfilmindex.ie/definition.

Dowden, Bradley. 2023. "Time | Internet Encyclopedia of Philosophy." Accessed December 9, 2023. https://iep.utm.edu/time/.

Egerer, Claudia. 2022. "Imaginings of Water: Anthropocene Waters and the Entanglement of the Living." In *Water Lore Practice, Place and Poetics*, edited by Camille Rouliere and Claudia Egerer. London: Routledge.

Ferwerda, Susanne. 2024. "Blue Humanities and the Color of Colonialism" In *Environmental Humanities* 16 (1): 1–18. https://doi.org/10.1215/22011919-10943081.

Heald, Karen. 2013. "In-between-Ness." Accessed March 22, 2025. http://www.karenheald.co.uk/research-in-between-ness.

Heald, Karen. 2023. *The Timekeeper and the Hour Glass: Artist Moving Image and Poetic Nonlinear Aspects of Time*. Film: 9 mins 20 secs. 16mm / Digital. https://www.karenheald.co.uk/films.

Jeans, James. 1931. *The Mysterious Universe*. Cambridge: Cambridge University Press.

Kristeva, Julia. 1979. *Women's Time*. Translated by Alice Jardine and Harry Blake. *Signs* 7 (1): 13–35.

Kristeva, Julia. 1980. *Desire in Language: A Semiotic Approach to Literature and Art*. Edited by Leon S. Roudiez. Translated by Thomas Gora, Alice Jardine and Leon S Roudiez. New York: Columbia University Press.

Kristeva, Julia. 1984. *Revolution in Poetic Language*. Translated by Margaret Waller. New York: Columbia University Press.

Kristeva, Julia. 2010. "Giotto's Joy." In *Philosophers on Art from Kant to the Postmodernists: A Critical Reader*, edited by Christopher Kul-Want. New York: Columbia University Press.

Loaiza, Estefanía Peñafiel and Gayatri Chakravorty Spivak. 2021. *Can the Subaltern Speak?* Köln: Walther König.

Minh-ha, Trinh T. (dir.). *Reassemblage: from the Firelight to the Screen*. 1982. Distributor: "New York Film Festival." 40 mins, 16mm transferred to digital file.

Minh-ha, Trinh T. (dir.). *Naked Spaces: Living is Round*. 1985. Distributor: "Women Make Movies." 137 mins, 16mm transferred to digital file.

Minh-ha, Trinh T. (dir.). *Surname Viet Given Name Nam*. 1989. Distributor: "Women Make Movies." 108 mins, 16mm transferred to digital file.

Minh-ha, Trinh T. (dir.). *Night Passage*. 2004. Distributor: "Women Make Movies." Digital film.

Minh-ha, Trinh T. 2005. *The Digital Film Event*. New York: Routledge.

Minh-ha, Trinh T. 2013. *D-Passage: The Digital Way*. Durham: Duke University Press.

Munteanu, Nina. 2018. "When Water Entangles." *The Meaning of Water* (blog), September 22: https://themeaningofwater.com/2018/09/22/when-water-entangles/.

Noë, Alva. 2023. *The Entanglement: How Art and Philosophy Make Us What We Are*. Oxford: Princeton University Press.

Phillipson, Michael. 1995. "Managing Tradition: The Plight of Aesthetic Practices and Their Analysis in a Technoscientific Culture." In *Visual Culture*, edited by Charles Jencks. London and New York: Routledge.

Radhakrishnan, Shivani. 2022. "Trinh T. Minh-ha." *Bomb* (interview), July 18: https://bombmagazine.org/articles/2022/07/18/trinh-t-minh-ha.

Steyerl, Hito. 2011. "In Free Fall: A Thought Experiment on Vertical Perspective." *E-flux Journal* 24, https://editor.e-flux-systems.com/files/67860_e-flux-journal-in-free-fall-a-thought-experiment-on-vertical-perspective.pdf.

Steyerl, Hito. 2012. *The Wretched of the* Screen. New York, Sternberg Press.

Spivak, Gayatri Chakravorty. 1988. "*Can the Subaltern Speak*?" In *Marxism and the Interpretation of Culture*, edited by Cary Nelson and Lawrence Grossberg. Macmillan Education: Basingstoke.

Tiwari, Ananya. 2017. "Julia Kristeva 'Women's Time' (Summary)—1979." *Linear vs Monumental Time* (blog), May 12: https://ananyatiwari.wordpress.com/2017/05/12/julia-kristeva-womens-time-summary-1979.

Toril, Moi. 1979. *The Kristeva Reader: Women's Time*. Oxford: Blackwell Publishing.

Walton, Millie. 2021. "Grasping the Eternal: Tacita Dean's 'fernsehturm'." *Trebuchet*, February 19: https://www.trebuchet-magazine.com/tacita-dean.

Williams, Adam. 2018. "Revolving House Takes its Interior for a Spin." *New Atlas*, July 10: https://newatlas.com/revolving-house-t-mochida-atsuko/55385/.

Yentob, Alan. 2018. "Tacita Dean: *Looking to See*." YouTube video, posted by lalulula TV, https://www.youtube.com/watch?v=nZc-OX5n9KI.

CHAPTER 15

Measuring Uncertainty

Emily DiCarlo

Abstract

This chapter investigates the standardization of Coordinated Universal Time (UTC) and the broader implications of 24/7 global timekeeping through the lens of art, science, labor, and history. It centers on *The Propagation of Uncertainty* (2020), a three-channel video installation filmed in Canada's official time lab, named after the statistical method used to calculate UTC—an equation that accounts for error in every value. Using the four uncertainties that underpin UTC as a conceptual framework, the chapter critiques the infrastructures and processes of time measurement, highlights marginalized temporal experiences, and shows how time-based art can challenge conventional notions of time by employing research-creation as an alternative, embodied mode of knowledge production.

Keywords

Coordinated Universal Time – contemporary art – post-real time process – labor standardization variable – *The Propagation of Uncertainty* – Bureau of Weights and Measures – *Circular T* – temporal resistance – video installation – time-based art – Taylorism – time-specificity – research-creation – radiation half-life – personal equation – mechanical objectivity – body-as-measuring device – the drift – *power-chronography* – caesium

1 Introduction

Coordinated Universal Time (UTC), the official time standard that airports, stock markets, and wireless communications rely on, is anything but absolute. These flows of exchange operate through UTC's finely-tuned temporal orchestration, a seemingly fixed and authoritatively omnipresent system. In actuality, the International Bureau for Weights and Measures (BIPM) in Paris, France, collects asynchronous data from 100 time laboratories around the world, each housing multiple caesium-driven atomic clocks, and then algorithmically

 | DOI:10.1163/9789004723634_017

sorts the localized time discrepancies through a month-long operation.[1] This process is routinely finalized through face-to-face negotiation, and once reckoned, BIPM sets the official universal time standard through the production of a report known as the *Circular T*.

The *Circular T* report works within a dynamic timeline, referred to as the "post-real time" process, to determine Coordinated Universal Time.[2] The BIPM first collects an asynchronous dataset from the 100 official time labs over a five-day interval, a process that takes anywhere from 10 to 45 days to complete. It then uses predictive algorithms to extrapolate the future's official clock time. Due to this lengthy turnaround, any current clock's time reflects gathered data from approximately one month ago. Thus, in reality, a multiplicity of deviated pasts informs our future's official, singular present.

Most apparent within the "post-real time" process is the range of unknown variability at every juncture. How do you measure uncertainty? In general, when developing a measurement model to accommodate the dynamic expressions of uncertainty, it must account for the fluctuating quantities related to the measured phenomena, the specific effects arising within the measurement process, and the interaction with the measured sample itself.[3] To produce the UTC standard, the BIPM calculates the uncertainty of each contributing time lab by considering a) the instability of the standard itself, b) the discrepancy between the conceived standard and the Earth's natural rotation, c) the interference occurring during communication processes, and d) the unknown effects of the system as a whole.[4]

Taking its name from the statistical calculation used in the *Circular T* report creation, *The Propagation of Uncertainty* (2020) is a three-channel video installation filmed at the National Research Council (NRC) in Ottawa—the sole Canadian contributor to the global UTC lab network. Using the four

---

1 As of February 8, 2022, 100 laboratories participated in creating UTC. The number of contributing sites is constantly in flux based on their atomic clocks' operational status and the addition or subtraction of BIPM Member and Associate States.

2 Dr. Patrizia Tavella, Department Director in the Time Department at BIPM, explained that the term "post-real time" process comes from a variation of *a posteriori*, which denotes "reasoning or knowledge which proceeds from observations or experiences to the deduction of probable causes." Patrizia Tavella, email message to Emily DiCarlo, "Graduate Student Research on UTC," February 17, 2020.

3 Joint Committee for Guides in Metrology (JCGM), "Guide to the Expression of Uncertainty in Measurement—Part 6: Developing and Using Measurement Models," *JCGM GUM-6:2020*. First edition, 2020.

4 Gianna Panfilo, "The Coordinated Universal Time," *IEEE Instrumentation & Measurement Magazine* (June 2016), 28–33.

uncertainties in the Coordinated Universal Time calculation as a conceptual framework, this chapter proposes *research-creation* as an alternative approach to critical time studies.[5] Weaving together examples from across disciplines, I will consider the materiality of time-measuring infrastructures and their processes, excavate accounts of marginalized temporal experience, and propose how art can engage with the variable nature of temporal experience to disrupt hegemonic notions of time.

2 The Propagation of Uncertainty

To create *The Propagation of Uncertainty* (2020), I partnered with the National Research Council (NRC) in Ottawa, the only contributing Canadian time lab to the Bureau of Weights and Measures's (BIPM) UTC atomic clock network. Working with Dr. Marina Gertvolf, the team lead at the Meteorology, Frequency & Time Department, I conducted multiple research trips, interviewed lab workers, toured the facilities, and carefully observed the material infrastructure of their time-keeping operations. Coming to these site-visits as an artist-researcher, as opposed to a fellow scientist, I was not interested so much in the mechanics as I was in the spatial poetics of this governmental space. What would time feel like in the place where the country's clock-time originated? During these visits, I saw very few workers. Instead, I noticed traces of human activity in the form of hand-jotted reminders and temporary DIY solutions scattered throughout the lab. Once granted permission to film in the NRC's time signal "dissemination room," a windowless laboratory, aesthetically unchanged since 1970, I embarked on a meticulous four-hour visual survey and site-specific performance-for-video, while the NRC's time-telling service—a 24-hour Talking Clock—narrated the unrelenting cadence of passing ticks over the room's P.A. speakers.[6] The production results in a 5-minute and 40-second

5 According to the Social Sciences and Humanities Research Council of Canada, *research-creation* is "an approach to research that combines creative and academic research practices, and supports the development of knowledge and innovation through artistic expression, scholarly investigation, and experimentation. The creation process is situated within the research activity and produces critically informed work in a variety of media (art forms)." See "Definition of Terms," Social Sciences and Humanities Research Council of Canada, accessed September 27, 2023. https://www.sshrc-crsh.gc.ca/funding-financement/programs-programmes/definitions-eng.aspx#a22.

6 The NRC's Talking Clock is one of their time-telling services and can be accessed through shortwave radio and a telephone hotline. The late Canadian news radio announcer Harry

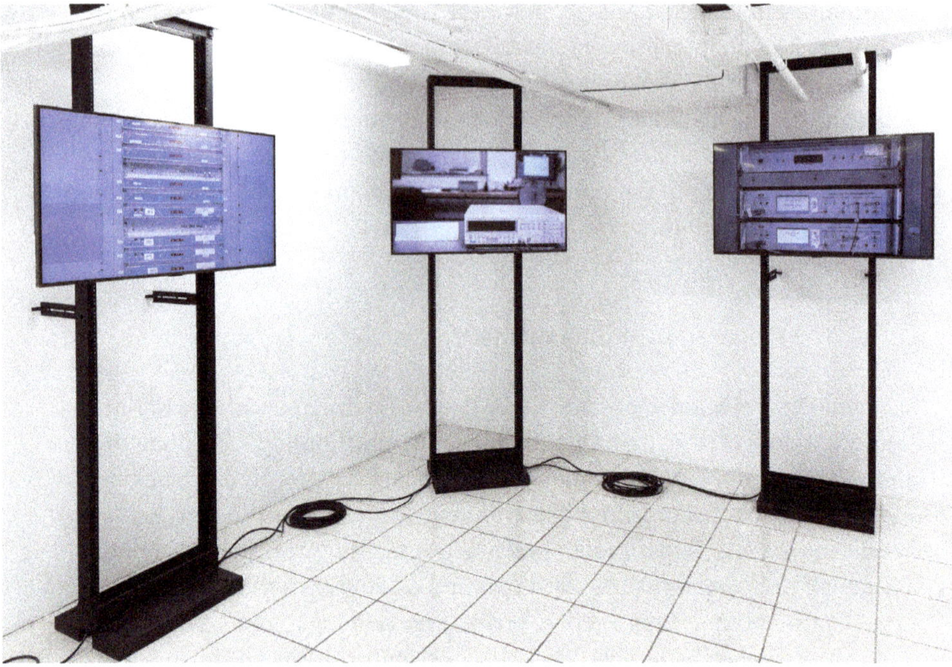

FIGURE 15.1 *The Propagation of Uncertainty*, 2020, three-channel video installation, 5 min. 40 sec. The plumb gallery, Toronto, Canada
PHOTO © ALISON POSTMA

video work that unfolds over three tv monitors, each installed on an industrial steel server rack, the same structures designed to organize networking equipment in data centers and, in this case, the time lab.

The Propagation of Uncertainty opens on the center screen, depicting an analogue clock face ticking away from 11:59:00 to noon while an authoritative male voice announces the time in 10-second intervals, followed by tone beeps to indicate the exact time. During this seemingly prolonged and anticipation-filled first minute, glimpses of the laboratory are revealed—a hand shifts a binder labelled "Time Lab"; a phone blinks "Message Waiting"; a hand-jotted note beside it reads "Talking Clock"; another hand pulls the

Mannis, who has been telling the time since 1960 on both services, still, rather ironically, tells the time from beyond the grave. Try it for yourself: +1-613-745-1576. See "Under the microscope: NRC's digital talking clocks," *National Research Council Canada LinkedIn*, November 20, 2023. https://www.linkedin.com/pulse/under-microscope-nrcs-digital-talking-clocks-6izse/?trk=article-ssr-frontend-pulse_more-articles_related-content-card.

FIGURE 15.2 *The Propagation of Uncertainty*, 2020, three-channel video installation, 5 min. 40 sec. Video still
PHOTO © EMILY DICARLO

"Emergency Procedures" folder from a shelf while a binder scrawled with the words "Servers 4 Clock Monitoring" falls sideways. The flanked T.V. screens oscillate back and forth on beat like a waving metronome while the center screen's clock ticks uninterruptedly. Nearing noon, an additional 17 talking clock voices from 13 time zones join the video's soundtrack, all announcing noon in their respective localities, surging into a single moment of discordant simultaneity.

The following sequence returns to the original single local voice, counting forward, "NRC, Eastern Standard Time, 12 hours, 0 minutes, and 10 seconds" while, in prolonged shots, a lab workspace frames a bank of drawers that sit patiently untouched until a hand opens one of them. Anticipation mounts for "something to happen." Eventually, sequenced voices from the climaxed cacophony are reintroduced in brief talking clock duets with the NRC voice, always anchored by the foregrounding of Eastern Standard Time (UTC-04). Moving chronologically by time zone, first the National Institute of Information and Communications in Washington, D.C., USA (UTC-04) duets with the NRC, then the "Time & Temp" hotline run by the Mercantile National Bank in Northwest Indiana, USA (UTC-05) takes its turn, followed by the telecommunications company SaskTel in Regina, Saskatchewan, CA (UTC-06),

FIGURE 15.3 *The Propagation of Uncertainty*, 2020, three-channel video installation, 5 min. 40 sec. Video still
PHOTO © EMILY DICARLO

and so on.[7] It is at this moment that a moving visual circuit begins, dancing left-to-right across the screens, briefly sharing additional glimpses of the time lab's workings: a hand-written sign reading "CALIBRATION IN PROGRESS," dated circuit boards with analogue dials and flickering lights, and collections of coaxial cable snaking through the space.

Then, the flanked T.V. monitors go black, and the center screen's light punctuates the viewing space with a series of zoomed-in, one-second flashes of blinking "1Hz" lights. Frequency, that is, the number of cycles passed through a fixed point in unit time, is measured in hertz, where a single hertz equals one second. The viewer thus not only experiences a representation of the time standard for the second, but is also effectively conditioned by the flashing one-second image intervals, which act as a calibrating force.

Still flashing, a sonic shift occurs in the soundtrack and for the first time, the unedited talking clock tone modulates and begins to drop in scale, signaling the video's final sequence. As the tone descends to a subterranean pulsing bass, socked feet from an undisclosed body lying on the laboratory's linoleum floor protrude from the left-hand frame. The image slowly pans right and a set of

7 The "Time Lady" service, as locals affectionately referred to it, was discontinued in 2023.

knees, legs, a waist, and hands resting on the torso of a jump-suited body come into the frame while laboratory equipment blinks behind. The bright blips and beeps of the field recordings heard earlier now ride on a darker current of deep bass. Just as the left screen's pan reaches the figure's neck and their feet fall out of frame, the center screen switches on, repeating the sequence, revealing the same socked feet, which lie motionless on the lab floor. As the left screen continues to pan right, and the figure's face is revealed, the body contracts into the center screen, eventually expanding again to create an elongated figure, mutated and distorted with four arms. The center image follows the sequence again and begins to pan, expanding and contracting. This editing strategy visually nods towards the repetition of clock-time while simultaneously butting up against the elasticity of experience.

Coined by neuroscientist David Eagleman, the subjective temporal experience of time "expanding," commonly associated with the experience of traumatic events, is known as *the oddball effect*. While time seems to dilate in these moments of disruption, the event awakens our senses, allowing us to process environmental stimuli at full capacity. It is not that duration expands;

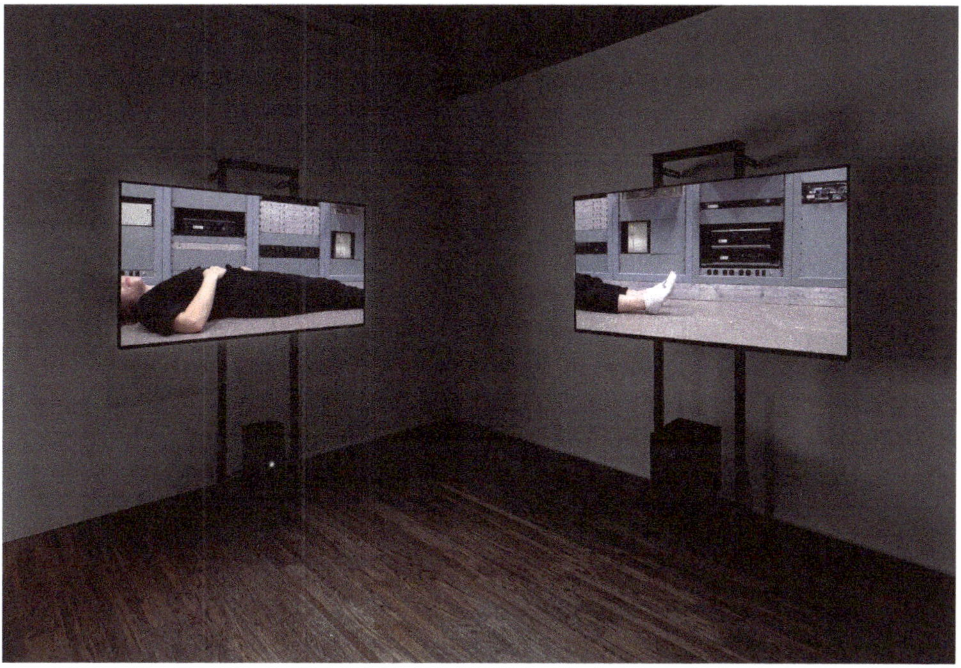

FIGURE 15.4 *The Propagation of Uncertainty*, 2020, three-channel video installation, 5 min. 40 sec. The Art Museum, Toronto, Canada
PHOTO © TONI HAFKENSCHEID

FIGURE 15.5 *The Propagation of Uncertainty*, 2020, single-channel video screening, 5 min. 40 sec. *New Toronto Works Festival*, Pleasure Dome, Toronto, Canada
PHOTO © EMILY DICARLO

rather our awareness of the regular-paced moment heightens.[8] This phenomenon of new, unexpected events perceptually stretching time, particularly well-known when linked to trauma, is best understood in contrast to "repetition suppression." Studies using electroencephalograms and PET scans have found that neurons become more efficient in processing repeated or familiar stimuli and that firing rates slow down as a sequence progresses. As the brain assumes that the series of experiences contains a pattern, it can conserve energy by switching to a form of perceptual autopilot, glossing over details already well-documented in memory.[9] *The Propagation of Uncertainty* takes advantage of these multiple registers by setting a cool, structuralist approach to time-keeping against the final moments of the documented performance,

8 In the weeks leading up to the filming of *The Propagation of Uncertainty*, a series of traumatic events occurred, including my closest friend's father dying tragically while on vacation. These events directly informed the performance-for-video's final gesture of laying on the floor, an act of refusal and simultaneous resolve. Death, the ungraspable horizon of our temporal experience, forefronted my thoughts while surveying the NRC's time lab as I anticipated attending the funeral hours after the production.

9 Alan Burdick, *Why Time Flies: A Mostly Scientific Investigation* (New York: Simon & Schuster, 2017), 156.

which convey a hyper-awareness of and presentness to the felt moment, where experiential time bloats to the brim of our attention.

In their jointly written guest exhibition text for the work's first installation at the Art Museum in Toronto Canada, Nadim Samman, Curator at KW Institute for Contemporary Art, Berlin and Dehlia Hannah, Associate Professor of Environmental Aesthetics at the University of Copenhagen, describe the effect in this way: "Towards the end of DiCarlo's video the camera pans, and you see the artist lying on the floor—her eyes open in what might be bliss, or boredom's release into a new thought. Lying on the linoleum floor, at the foot of a bank of aluminum circuit boards, wearing a black worker's suit, in the midst of the room's crushing inhumanity, her face is illuminated by a glow. You know it issues from a fluorescent light, but somehow DiCarlo's expression—upward turned, mouth slightly open—suggests some form of beatitude, like she is being cradled by time, resting in its lap or belly. It is at this point (towards the end of the video) that you, the viewer, have finally slowed down enough to watch it again—at its own pace. Now, there is something deeply comforting about the speaking clock [...] Looking at her, she appears to have overcome any disjunction between her intimate temporality and that of Ottawa's national timekeeping laboratory. Where does this leave the rest of us? Her silent figure refuses to say."

3 The Instability of the Standard Itself

Objectivity requires an unbiased, non-emotional perspective based on methodological protocols, which, in turn, as a corrective strategy, suppress some aspects of the self.[10] During the 19th century, the idea of an objective *scientific self* was built up and reinforced through laboratory practices that minimized the impact of personal will. Specialized self-disciplinary techniques, such as grid-guided drawings, real-time entries in lab notebooks, and later, the camera, contributed to a false division of the self into an active experimenter and a passive observer. Reinforcing this notion, key positivist thinkers such as Auguste Comte (1798–1857) and later Karl Popper (1902–1994) believed that meaning existed independently of feelings, beliefs and interpretation, asserting that true knowledge only arose from perfectly reproducible experiments. While 19th century scientists strove to deny subjectivity at all costs, artists of

10 Lorraine Daston and Peter Galison, *Objectivity* (New York: Zone Books, 2007) 36.

the same period, ironically, were encouraged to embrace their individuality through movements such as Impressionism, Romanticism and Art Nouveau.

The desire for objectivity and scientific reproducibility, coupled with the accelerated production, trade, travel and communication of the Industrial Revolution, birthed the concept of standardization. Standardization—the process of aspiring and conforming to a regulated ideal—was interlinked with the higher pursuit of objective truth. Through controlled consistency, the cult of measurement promised predictability, an assured knowingness of the world, and a sense of power over the sublime. In order to facilitate and officiate standardized measurement systems on a global scale, a coordinated, international governing body was needed, so in 1875, at the seminal Metre Convention in Paris, France, The Bureau of Weights and Measures (BIPM) was established. The organizing body, currently comprised of 64 Member States and 36 Associate States and Economies,[11] now defines and governs the International System of Units (SI). It is the only measurement system, used in every country in the world, which includes the seven foundational measuring units of length (meter), mass (kilogram), electric current (ampere), thermodynamic temperature (kelvin), amount of substance (mole), luminous intensity (candela), and, of course, time (second).[12]

Before establishing the BIPM, length-based measurements had been based on human proportions. In ancient Egypt, for example, the "royal cubit," or the length of the Pharoah's arm, elbow to fingertip, was the standard unit of measurement. To standardize this body-based unit, Egyptians produced corresponding granite rods which were used as measurement tools for construction. Of course, as the standard changed with each new incoming Pharoah, the measuring rods required continual updating. Moving from Egypt to Greece and then Rome, this body-based approach of standardization eventually found its way to Europe, where the length of a man's foot or walked stride defined a meter or yard. "For example, the pace—one left step plus one right step—is

11 See "Member States, and Associate States and Economies," BIPM, accessed November 3, 2023, https://www.bipm.org/en/-/list-of-member-states-associates.

12 When it comes to time-keeping and the contribution of data to the BIPM, there is no difference between "Member States" and "Associate States and Economies." The status differential is based on a State's paid membership contribution and experience. Contributing sites often begin as Associate States as an introductory status, eventually being encouraged by BIPM to ascend to full Member State status after five years. See "Explanatory note: Calculation of the individual subscriptions for Associates (States and Economies) of the General Conference on Weights and Measures (CGPM)," Bureau International des Poids et Mesures, January 2024, https://www.bipm.org/documents/20126/33107612/calculating-subscriptions-EN.pdf.

approximately a meter or yard. (On the other hand, the yard did not derive from a pace but from, among other things, the length of King Henry I of England's outstretched arm.) *Mille passus* in Latin, or 1,000 paces, is where the English word 'mile' comes from. However, the Roman mile was not quite as long as the modern version."[13] This variability continued through to the Middle Ages, when farmers measured the area of their land according to the time it took for a man, or a man with an ox, to plough in a day—a measurement that would shift depending on the labor intensity of harvesting different crops.[14] Interestingly, time defined the spatial metric through the body's pace and cumulative duration.

The accelerating speed of manufacturing during the Industrial Revolution gave rise to scientific management theory or "Taylorism." Considered one of the first management consultants, Frederick W. Taylor published his treatise "Shop Management" in 1903 in the journal, Transactions of the American Society of Mechanical Engineers, where he outlined how to increase efficiency and speed in manufacturing businesses, specifically in the steel industry, through "enforced standardization of all details and conditions, large and small, tools, appliances and methods" so that daily tasks could be "carried out with certainty." This system broke down factory labor tasks and timed the individual motions of piece work to determine the most efficient sequence and ideal working speed. In doing so, it also rendered a worker's body into a machine-like unit of measurement, conforming to a time-based standard founded on capitalist pursuits.

Taylor proposed that every worker should be measured against "the quickest time in which it can be done by a first-class man" and that "this time may be called the 'quickest time', or the 'standard time' for the job." He goes on to say that "the greatest obstacle to the attainment of this standard is the slow pace which they adopt, or the loafing or 'soldiering', marking time, as it is called." This *marking of time* he defines in two ways: *natural soldiering*, which is the instinctual tendency to "take it easy" and "work at a slow, easy gait" and *systemic soldiering*, which is the influence of group dynamics, where if the majority of workers move slowly, everyone else follows suit.

Taylor's obsession with standardization throughout the study seems morally based. He describes man's laziness as "the greatest evil from which both workmen and employers are suffering," emphasizing the hierarchy between those who pushed working speeds to their limits and those who moved at a

13 "SI Redefinition: Meter," National Institute of Standards and Technology, U.S. Department of Commerce, June 2, 2021, https://www.nist.gov/si-redefinition/meter/.

14 "SI Redefinition: Meter."

slower, natural pace. When considering workers' adoption of the new time standard, he concludes that the lazy worker would also "use every effort to prevent it and lead management in the wrong direction." Disregarding the reality of harsh work conditions and long hours in factories as reasons why laborers were unmotivated, in 1884, Taylor proposed that monetary incentivization, specifically, a differential rate for piecework, was the only way to encourage workers to move faster.[15] He believed that encouraged in this way, their material need for survival would drive workers to meet the new standard time more consistently.

Speed, expansion and progress characterized the 19th century, annihilating space and time through the rapid development of the railway system and telegraph transmission line networks throughout Canada, the United States, and Europe. Only a year prior to Taylor's time-labor incentivization program, Sir Sandford Fleming, Canada's chief railway construction engineer, used the International Meridian Conference in Washington, D.C. to propose dividing the world into 24 time zones, each 15 degrees of longitude apart, and using the Greenwich Meridian (GMT) as the prime meridian. Such proposals responded to the need for efficient, synchronized coordination, which was becoming integral to the accelerating flow of goods. To return to Taylor, laboring bodies were expected to keep up. Taylor's scientific methods to study work processes and further improve efficiency were designed to establish the new "standard time" for any factory job. Soon thereafter, "efficiency engineers" photographed and filmed "first-class men," scrutinizing frame-by-frame and plotting individual movements in units of a thousandth of a minute in order to edit out non-essential actions.[16]

Following in Taylor's footsteps, the US-based Methods Engineering Council developed the Methods-Time Measurement (MTM) system during the 1940s, which leveraged 16 mm film cameras to record factory workers. In her essay, "Measures of Time: Countering the Clock, Troubling the Timeline," curator Johanna Thorell discusses the role that film and photography played in the regulation of workers' bodies, pointing out that, "Not only were records of workers compared to a normative measure of time spent on specific tasks, but the camera itself performed as a "clock for seeing," by means of which manual

15 In 1890, the average steel worker made 10 dollars per week, which was barely above the poverty line of 500 dollars per year. See "The Steel Business," *Andrew Carnegie: The Richest Man in the World*, PBS, January 20, 1997, https://www.pbs.org/wgbh/americanexperience/features/carnegie-steel-business/.

16 Richard Salame, "The New Taylorism," Jacobin, February 20, 2018, https://jacobin.com/2018/02/amazon-wristband-surveillance-scientific-management.

labor could be captured, divided into minimal components, and studied to be rendered more time-efficient."[17] A reference to Roland Barthes, Thorell's use of a "clock for seeing" draws a likeness between a camera's ability to mechanically capture an objective representation of reality and a clock's ability to do the same.[18]

The role of surveillant technology within the temporal discipline of labor has only intensified during the past century. Referring to four of the nine movements of the Methods-Time Measurement (MTM) system, artist Kajsa Dahlberg's film, *Reach, Grasp, Move, Position, Apply Force* (2015) employs the equipment and services used in her film production as a departure point, from which to investigate the labor conditions under which they were made. The 40-minute film weaves archival footage of MTM film tests with interviews and testimonies from factory and warehouse workers at Apple, Panasonic and Amazon, alongside gig workers like the freelance translators employed to subtitle the film. In a self-reflexive instance, a representative from the Nordic MTM Association in Västerås, Sweden demonstrates, using the artist's video camera as it records, how they codify every micro-movement into an alpha-numerical data system, where one Time Measurement Unit (TMU) equals 0.36 seconds. We watch him break down the steps, "stretch," "grasp," and "apply force" (a concentration of power to release the mirror), and then analyze the increments of time to adjust for optimization and further control—the pervading theme in a film that exposes the circumstances of its own production.

In his book, *The Warehouse: Workers and Robots at Amazon*, researcher Alessandro Delfanti details the punishing rhythms warehouse workers endure during their 10-hour order-picking shifts in Amazon fulfillment centers. To close the temporal gap between seller and buyer, fulfilling Amazon's same-day, and sometimes two-hour, delivery guarantee, workers are constantly surveilled by an advanced "clock for seeing." The workers describe how "a pervasive surveillance system monitors their productivity at every step. The valuable information generated by their labor is captured and monopolized by Amazon's software systems, and then fed to the machines that run the warehouse. …"[19] Drawing from Taylorism's principle of the "standard time" set by the factory's fastest worker, the algorithm determines what is colloquially known

17 Johanna Thorell, "Measures of Time: Countering the Clock, Troubling the Timeline," in *Clocking Out: Time Beyond Measurement* (New York: Whitney Museum of American Art, 2023), 46.

18 See Roland Barthes, *Camera Lucida. Reflections on Photography* (New York, N.Y.: Hill and Wang, 1982), 15.

19 Alessandro Delfanti, *The Warehouse: Workers and Robots at Amazon* (London: Pluto Press, 2021), 10.

FIGURE 15.6 Kajsa Dahlberg, *Reach, Grasp, Move, Position, Apply Force*, 2015. HD video, 40 minutes. Video still

as the "Amazon pace" in order to extract as much work as possible from its employees.[20] A former worker describes the race against time, saying that the barcode scanner they carry dictates their tasks and "shows a countdown for each individual piece, for instance, a minute and a half to go get this item. It's a bar that dwindles. And [the pick tower] is gigantic. You need to walk for 4 minutes to get anywhere."[21] The worker exists only in discrete units of the present, with no time to think about the past and no foresight into the next item to be picked, stuck in a perpetual mind-numbing dash. The scanner also registers if a worker falls behind on the dictated pace. Should this happen too often, the worker will be pulled aside for reprimand. Turnover is high, as those who cannot keep up with the market's accelerating demands are quickly replaced.

The reality of working in an Amazon fulfilment center lays bare *the instability of the standard itself* when the standard comes to exceed physical human limits. When laborers can no longer keep up with the accelerating pace of consumer demands and corporate greed, Amazon's solution is to remove people

20 Delfanti, 43.

21 Delfanti, 42.

from the equation altogether. First tested in Italy in 2019 and now present in many of their fulfilment centers worldwide, the CartonWrap 1000 is a machine that can pack an order in three seconds—a time standard well beyond human capability. To keep the "Amazon pace" (and continually exceed it), the future of global supply chains will include "lights out" warehouses, where machines like the CartonWrap1000 work in complete darkness, also saving the company money on energy costs.[22] As technological automation redefines workplace time-motion standards and increasingly replaces human workers, the drive for speed and efficiency pushes time standards beyond human perception. At what point does the entire system itself buckle under these pressures?[23]

4 The Drift (the Discrepancy between the Conceived Standard and the Natural Rotation)

In their book *Objectivity*, Lorraine Daston and Peter Galison trace the evolution of objectivity through a lens of variability, subjectivity and omissions found in scientific visual history. Beginning in the 16th century, scientists customarily prepared visual surveys of their research in the form of image-based atlases as a way "to standardize the observing subjects and observed objects of the discipline by idiocrasies—not only those of individual observers but also those of individual phenomena."[24] From anatomical structures to microscopic crystals, scientists established a precedent for including only "perfect" representations as worthy categorical examples. The prevailing belief was that future science did not need a repertoire of documents with optical artifacts, biological abnormalities, or defective anomalies, such as asymmetrical snowflakes. Daston and Galison argue that the scientific pictorial taxonomies of this time are the result of consequential choices regarding knowledge, persona, and collective vision.

---

22 Delfanti, 120.

23 In July 2024, a faulty routine update to CrowdStrike's Falcon cybersecurity software led to a global IT outage, affecting approximately 8.5 million Windows systems. The affected sensor, which uses real-time machine learning to detect suspicious activity, crashed many critical sector systems, including airlines, banks, hospitals, and retail services. Despite the quick response, some systems required manual intervention to resolve the problem, sometimes taking weeks, depending on available resources. Considered an industry standard, used by multinational companies, the event exposed the precarious nature of our reliance on automated "black box" global time-critical systems. See Matt O'Brien, "Tech disruptions sparked by software update highlight the fragility of globally connected technology," *AP News*, July 19, 2024, https://apnews.com/article/tech-outage-crowdstrike-microsoft-windows-7294e6c9356050cb095dad3353e9244c.

24 Lorraine Daston and Peter Galison, *Objectivity* (New York: Zone Books, 2007), 63.

For 300 years, standardizing a myriad of natural phenomena into archetypal categories was the norm. Beginning in 1875, British physicist Arthur Worthington dedicated his life to creating a sequenced visual record of the complex process of fluid flow. Painstakingly, he would splash mercury or milk droplets on a hard surface, trigger a bright flash which would impress an afterimage into his retina, from which he would hand draw the shape immediately. For decades, Worthington edited out forms that did not fit the visual narrative he aimed to convey. In 1894, however, he captured a droplet's splash by photograph for the first time. Following this ground-breaking moment in observational science, he announced that he would discard all previous representations, proposing that visual archives must embrace all accidents and asymmetries to reflect the full complexity of nature.[25]

Both scientists and artists viewed the camera as a tool to capture essential truths, while simultaneously appreciating the invention's ability to register complex details of the world. However, where scientists lauded photography for its potential to achieve mechanical objectivity by producing automatic images free of human interpretation, artists scoffed at this reductive application, arguing that it disregarded the makers' individuality and imaginative interpretation of the world.[26] A distinction arose: mechanical photography facilitated the pursuit of scientific knowledge by producing an objective representation of the world, while aesthetic photography produced an artistic record of the essence of experience through portraiture and landscape.

In 1878, commercial photographer Eadweard Muybridge blurred the line between scientific documentation and artistic presentation when he famously used multiple cameras to capture a galloping horse's gait, forever putting to rest the debate of whether or not all four hooves ever leave the ground simultaneously.[27] As one of the earliest visual records of time-motion studies, his empirical evidence laid the groundwork for the rise of cinema and the moving image a decade later. Muybridge's record of sequential motion, captured at 1⁄500th of a second, rendered visible what had before been temporally imperceptible to the human eye.[28]

---

25 Daston and Galison, *Objectivity*, 16.

26 Romanticism in art and literature, spanning the late 18th to mid-19th centuries, emphasized emotional expression, sublimity in nature, mythology and mysticism—topics that lay beyond the bounds of scientific objectivity.

27 Regardless of his pioneering advancements across science, art and technology, public controversy ensued when he revealed that he had retouched the photos, a practice he did regularly as a commercial photographer. See Daston and Galison, *Objectivity*, 133.

28 Jimena Canales, *A Tenth of a Second: A History* (Chicago: The University of Chicago Press, 2009), 3.

The speed of thought, that is, the time it takes for a message to reach the brain, is a tenth of a second. Hermann von Helmholtz, a 19th century physicist and philosopher interested in the limits of perception, ascertained that "self-consciousness" lagged "behind the present" by a "tenth part of a second." In observational science, quantifying the perceptual moment was key to designing experiments that could remedy the *personal equation*. This phenomenon, first identified by astronomers, referred to discrepancies in observers' star-transit time calculations, caused by the delay between witnessing the event and responding to it. In attempts to establish a unit of measurement for the perceptual moment, scientists acknowledged the variability of cognitive processing, and admitted that the personal equation is rarely a precise tenth of a second, instead oscillating between one to a few tenths of a second.[29]

The *personal equation*, or the inherent variability of human experience, challenged the fundamental scientific operation of measurement.[30] Psychologist and philosopher William James describes the approximate measure of the durational moment as what he termed the "specious present." James says "The practically cognized present is no knife-edge, but a saddle-back, with a certain breadth of its own on which we sit perched, and from which we look in two directions in time. The unit of composition of our perception of time is a duration, with a bow and stern, as it were [...] We seem to feel the interval of time as a whole, with its two ends embedded in it."[31] For James, "the moment" is a slippery, dynamic unit of measurement rooted in feeling, whose temporal length is as long as the mind can hold it.

By the turn of the 20th century, variation was understood as a fundamental reality that permeated all inner experience and natural phenomena. Even the consistency of the 24-hour day was put into question when astronomer Maurice Lecornu discovered fluctuations in the Earth's rotating speed. His findings later revealed a deceleration pattern occurring every few decades, during which the days lengthen by a few milliseconds over several years before returning to an average rate. Since 1972, the variability of a day's length, or Earth's rotation time, has been measured against atomic clock time; when the difference becomes too large, a leap second is added to reconcile the two. In addition to reckoning with this discrepancy, timekeepers must also contend with the variations of the measuring instruments themselves when manufacturing Coordinated Universal Time.

29 Canales, 6.

30 Canales, 11.

31 William James, *The Principles of Psychology* (New York: H. Holt, 1890), 610.

Since 1967, scientists have addressed these problems of variation by relying on caesium's predictable oscillating frequency to define the global time standard. Despite being incredibly stable, however, a caesium atom can also experience minor deviation due to changes in temperature, pressure, and quantum uncertainties, causing the atom's frequency to fall out of tempo with the UTC standard, known colloquially as the clock's "drift."[32] As an ongoing protocol, therefore, timekeepers must use corrective action to re-synchronize their clocks' frequency to the standard.[33] Informally known as *tuning* in the time-keeping community, scientists employ a process of *syntonization* to carefully adjust the atom's oscillations in fine increments until it resonates again with that of a reference.

What if desynchronizing from the standard and embracing variability could lead to new knowledge production? In 1962, French geologist Michel Siffre did just that and pioneered the field of human chronobiology by using his body as a measuring instrument. During a 63-day research expedition, designed to chart a glacier in the pitch-dark Scarasson Cavern of southern France, he decided to "live like an animal, without a watch, in the dark, without knowing the time." Without regular daylight-night cues, his circadian rhythms gradually drifted from the standard 24-hour cycle to one lasting 24 hours and 30 minutes, uncovering an internal timing system that operates independently of natural day-night cues.

Using his body as measuring device, Siffre established a simple scientific protocol, where he would call his stationed team aboveground on a one-way telephone line upon waking, before eating, and just before sleeping. During these calls, he conducted two tests: taking his pulse and, as a psychological test, counting from 1 to 120 at a rate of one digit per second. Later in the experiment, it took him five minutes to count to 120, showing that he now perceived

32 Expressed as "UTC(k)" in the *Circular T* report, the "local realisation offset" represents a time laboratory's deviation in nanoseconds from the Coordinated Universal Time (UTC) world standard. Each local site is characterized by its error (k) relative to UTC. For instance, data from October 13, 2019, shows that the degree of error ranged from +0.1 nanoseconds at the European Space Technology Centre in Noordwijk, Netherlands, to +13,105.7 nanoseconds at the Laboratoire de Dynamique, Physique et Modélisation des Réseaux in Tunis, Tunisia. See The International Bureau of Weights and Measures, *Circular T: no. 382* (November 7, 2019, 15h UTC). France, Paris, 2019. Accessed January 30, 2020. https://www.bipm.org/en/time-ftp/circular-t.

33 Dr. Masami Yasuda, Group Leader for Time Standards at the Research Institute for Physical Measurement and National Metrology Institute of Japan, poetically likened the continuous process of steering the frequency drift back on course to that of "driving a heavy ship." Dr. Masami Yasuda, in-person interview with Emily DiCarlo, Advanced Industrial Science and Technology (AIST) in Tsukuba, Japan, July 19, 2023.

five conventional minutes as though they were two. Subsequent experiments with other research subjects demonstrated that many of them achieved a 48-hour sleep–wake cycle, sustaining 36 hours of continuous activity followed by 12–14 hours of sleep.[34] And in 1972, Siffre descended again underground in Del Rio, Texas, this time for 205 days, and finally experienced his own 48-hour sleep–wake cycle.[35] Upon resurfacing, he reflected on the psychological effects of the temporal drift, describing the experience as like inhabiting a seemingly eternal present, saying "It's like one long day. The only things that change are when you wake up and when you go to bed. Besides that, it's entirely black."

Examined through visual representation and auditory effect, *The Propagation of Uncertainty* explores the "drift" of time as a desynchronizing process in which experiential felt time falls out of tempo with the regime of clock time. The opening sequence of the video installation primes the viewer with a steady visual survey of the lab's environment, backtracked by a constant rhythm of layered talking clock recordings. During the final half, the body-as-measuring-device, now in repose, expands and contracts across the three T.V. screens, and a gradual rhythmic shift signals the transition from regimented measured clock-time to experiential durational time.

Dr. Martin Scheuregger, Senior Lecturer in Music in the School of Creative Arts at the University of Lincoln (UK), analyzes the change of background sound in *The Propagation of Uncertainty*, saying, "Just over halfway through the work, the temporal patterns start to become disrupted with changes clustering around the temporal golden section (3:30 of the total 5:40 duration) [...] at 4:10 a strong bass pattern is introduced that reframes the temporal grid, its driving rhythm moving the first beat of the measure by half a second, so that we begin to hear the ticking clock as an off-beat or, more accurately, as a backbeat. The position of this backbeat also suggests a doubling of the pulse—rather than counting beats at 60 bpm, we move to 120 bpm, with the ticking on beats two and four. From here to the 5:20 mark, there is an interplay between the lilting five-beat voice pattern at 60 bpm, and the double-time four-four of the bass and ticking clock "drum beat." In the final twenty seconds, only the bass remains, the four-beat pattern seemingly winning out. This shift from a lilting, familiar voice to a more menacing, upbeat and directly musical drumbeat suggests the commodification or capturing of time, or even its colonization."

34 This discovery piqued French military interests, especially as they had just initiated their nuclear submarine program and were interested in extending soldiers' wakeful operations.

35 NASA funded Siffre's second major research trip to study the psychological well-being of crews on space expeditions.

Despite the reposed figure's resistance, 24/7 standardized time eventually envelopes all.

In addition to the tempo change, a subtle pitch shift occurs in the background of the piece, facilitated by introducing a constant binaural tone. An effect particularly pronounced when experienced through headphones, two slightly different frequencies play through the left and right-side speakers, resulting in the listener perceiving a third frequency not present in the original sounds. Despite the standard beat pushing forward, the body-as-measuring device drifts screen-to-screen while the sound's frequency slowly rises and falls, resisting synchronization with the original pitch.

5 The Interference in the Process

Interference refers to phenomena in which external influences disrupt the performance or accuracy of a system. In global timekeeping, this interference manifests as *noise*—unwanted and often unknown variances that the measuring signal undergoes during capture, storage, transmission, processing, or conversion. When determining Coordinated Universal Time (UTC), the uncertainty of *jitters*—or rapid variations in clock signal intervals caused by electromagnetic waves, environmental conditions, and signal processing issues during satellite data transfers between time labs and BIPM—can disrupt synchronization with the official time standard.

The social conditions of UTC—the always-on, finely tuned global standard that drives a neoliberal, technocratic society—has led to what cultural theorist Sarah Sharma calls *power-chronography*, or the temporal architecture that shapes labor relations. She argues that experiential time is unevenly distributed depending on our socio-economic standing and corresponding systemic barriers. For those working night-shifts, such as janitorial custodians, healthcare professionals and taxi drivers, working time is dictated by and synchronized to the temporal demands of regular daytime working populations. Similarly, those who work in offshore call centers are forced "to live within the tempo of business hours elsewhere, subordinated to the time demands of another population."[36] Necessary to the continual function and flow of 24/7 operations, the demands for temporal coordination often render these individuals invisible, relegating them to the fringes of society, and resulting in "the experience

36 Sarah Sharma, *In the Meantime: Temporality and Cultural Politics* (Durham: Duke University Press, 2014), 165.

of laboring within a temporal infrastructure while being cast outside of it."[37] Within the political economy of time, Sharma advocates for a radical and democratic political approach to rebalancing our time-space distribution.

By reimagining time as a collective struggle against temporal exploitation, we are encouraged to embody the *noise, jitters*, or *interference* of global timekeeping as acts of resistance, disrupting the relentless grind of standardized time regimes. Cultural theorist Jonathan Crary proposes that "the imaginings of a future without capitalism begin as dreams of sleep. These would be intimations of sleep as radical interruption, as a refusal of the unsparing weight of our global present."[38] *Intimations of sleep*, like daydreaming, resting and actively doing nothing, manifest as non-productive actions that serve as a conscious disruption into the grinding demands of globalized time. When reframed this way, these acts of resistance protest the pressures of unrealistic sleep-wake rhythms and reject the digital appeals for our captured attention, making it possible to engage in imaginative world-building and rehearse alternatives to our current temporal predicament.

Filmed using a hidden camera, Finnish artist Pilvi Takala documents her undercover performance as a new hire in Deloitte's marketing department in *The Trainee* (2008). Over the course of a month, she conducts simple interventions: sitting in an open-concept workspace without a computer or notebook, claiming to do only "brain work"; staring into space for hours in the tax department library; and riding the elevator endlessly while explaining to her bewildered co-workers that she planned to ride it all day and that "the rhythm of thought is different here." The final sequence of *The Propagation of Uncertainty* also explores the body as a site of resistance through performance, while the camera, or "the clock for seeing," bears witness to an act of refusal.

6 The Unknown Effects of the System as a Whole

As previously discussed, global time-keeping involves many subsystems riddled with uncertainties, including factors such as the stability of the standard, the calibration process, human variability, instrumental drift, and data processing interference. When these uncertainties are aggregated to consider the system as a whole, the unknowns proliferate exponentially, revealing a profound precariousness inherent in the entire structure.

37 Sharma, 57.

38 Jonathan Crary, *24/7: Late Capitalism and the Ends of Sleep* (New York: Verso, 2013), 128.

FIGURE 15.7 Pilvi Takala, *The Trainee*, 2008. Installation. Three videos with a duration of 13:52, PowerPoint presentation, keycard, letter. Video still. Courtesy of the artist and Carlos/Ishikawa, London. HD video, 40 minutes
COURTESY OF THE ARTIST AND VIDEO-FORUM OF NEUER BERLINER KUNSTVEREIN, BERLIN, GERMANY

What happens if the system itself collapses? Or what happens if a time lab experiences a blackout and goes offline? What are the unknown, short-term material effects and long-term repercussions? In a rare case study, the National Institute of Information and Communications Technology (NICT) in Koganei, Tokyo, and the National Institute of Advanced Industrial Science and Technology (AIST) headquarters in Tsukuba, Japan—who work in tandem to continuously disseminate Japan Standard Time and maintain a "high precision space-time infrastructure and frequency standard"—sustained complete system failure on March 11, 2011.[39] A massive earthquake off the northeast coast of Japan put into motion a tsunami with 130-foot waves, which flooded the Fukushima Daiichi Nuclear Power Station and caused three reactors to

39 National Metrology Institute of Japan (NMIJ) and National Institute of Advanced Industrial Science and Technology (AIST), "The current status of Acoustics, Ultrasound and Vibration measurement standards at NMIJ/AIST," BIPM, accessed March 20, 2020, https://www.bipm.org/documents/20126/28435028/working-document-ID-5165/97806bc9-7aac-6958-1f5e-96e33c19c0da.

meltdown in a level seven catastrophe (on a scale of one to seven), killing more than 20,000 people and displacing 160,000 more.

Generally, routine system safety measures suffice, such as running three simultaneously operating servers backed up by large UPSs or "uninterruptible power supplies" that seamlessly provide battery-stored energy during a power outage. As a final cautionary layer of protection, should all else fail, the primary devices connect to backup generators, which have enough fuel to maintain power for three days.[40] In this case, however, due to seismic activity damaging measurement systems at the NICT, and electrical power shortages at the AIST, operations came to a halt.[41] Japan broadcasts its time signal from two radio transmission stations, one on either side of the country. Located 20 kilometers from the Fukushima Daiichi power plant, the damaged Ohtakadoya-yama station suspended all operations for 40 days, resuming its broadcast only five months later when the NICT made infrastructural changes to support remote operations.[42] This unexpected event forced the Eastern region to "borrow time" from the Hagane-yama station on the west coast using NTP (Network Time Protocol) and GNSS (Global Navigation Satellite System).[43] However, as the intensity of radio waves exponentially decreases with distance, this left dead zones on the out-of-reach North-East coast. Two years later, as a result, the NICT shifted strategies and created a decentralized power system to generate

40 National Institute of Information and Communications Technology, "Report of Time and Frequency Activities at NICT (2012–2015)," accessed June 3, 2021, https://www.bipm.org/documents/20126/28435182/working-document-ID-5264/03fbf334-13b1-9a83-db1d-296919adfbda. Additionally, Dr. Marina Gertsvolf, at the Time and Frequency Department, National Research Council (NRC) in Ottawa, Canada, corroborated the referenced system safety measures as a standard precaution. In our email correspondence on March 30, 2020, she cited the NRC's hydro station explosion on Boxing Day in 2016, which knocked out all of the campus' power, causing emergency generators to take over successfully. Because of the redundancies in place, the production of time data was not interrupted in Canada.

41 Reported damage included all six caesium atomic clocks and four hydrogen masers at the noted facilities. See National Metrology Institute of Japan (NMIJ)/AIST, "19th Meeting of CCTF, Research Activities on Time and Frequency," BIPM, accessed March 20, 2020, https://www.bipm.org/documents/20126/48885527/working-document-ID-9453/a53de2ff-3bf9-ed8b-d726-bcbacd927a8d.

42 National Institute of Information and Communications Technology, "Japan Standard Time—history," accessed April 30, 2024, https://www.nict.go.jp/en/sts/history.html.

43 In the event that a country loses its official site and its ability to produce time data, the country will draw the necessary information to keep time from another country's site, using NTP (Network Time Protocol) and GNSS (Global Navigation Satellite System). Confirmed by Dr. Marina Gertsvolf (Team Lead, Time and Frequency Department, National Research Council) in discussion with the author, March 30, 2020.

and disseminate Japan Standard Time, which involved constructing a substation in Kobe, Japan, housing five caesium atomic clocks.[44]

Deemed a human-made disaster by both the Japanese government and an independent investigation, the Fukushima Exclusion Zone, or what is locally known as the "Difficult-to-Return" zone, includes seven contaminated municipalities, totaling 337 square kilometers as of March 2021. Initially evacuated within a twenty-kilometer radius of the plant, the government later expanded the zone to follow air currents, which picked up and spread plumes of radioactive material, consisting mainly of the unbalanced element caesium-137, to as far as Tokyo.[45] Given that the International Bureau of Weights and Measures has, since 1967, defined the second as a unit of time by the frequency of caesium-133's ground state transition, it is ironic that caesium, known for its atomic predictability, is now, in its unstable form, the predominant nuclear by-product contaminating the soil, water, and environment of the Fukushima region.

Radioactive materials, among the longest-lasting human-made substances, can persist for decades or even thousands of years. Notably, caesium-137 exhibits a physical half-life decay of 30 years, gradually reducing the quantity of radioisotopes present in the environment and eventually decreasing the risk of cancer. This temporal framework proved foundational to artists Nikolaus Hirsch and Jorge Otero-Pailos' ongoing site-specific project, *Becoming Monument* (2015–present), as part of the "Don't Follow the Wind" exhibition, curated by the collective of the same name, Don't Follow the Wind (Chim↑!Pom from Smappa!Group [initiators], Jason Waite, Kenji Kubota, Eva and Franco Mattes). Situated within the inaccessible Fukushima Exclusion Zone, the exhibition features twelve commissioned artworks that "act as material witness to the contamination and the durational separation of former residents."[46] Influenced by the Conceptual Art and Land Art movements of the 1960s and

44 National Institute of Information and Communications Technology, "Decentralized Japan Standard Time: Time-link between separated facilities," accessed April 30, 2024, https://www.nict.go.jp/en/sts/decentralized_jst.html.

45 Nikolaus Hirsch and Jason Waite, eds., "Don't Follow the Wind," *Critical Spatial Practice* 12 (London: Sternberg Press, 2021), 15.

46 As the twelve site-specific artworks located within the Fukushima Exclusion Zone remained inaccessible for in-person viewing, the curators developed a series of satellite "Non-Visitor Centers." These centers, showcasing video, photographic, and 360-degree VR documentation, have been integrated into exhibitions at The Watari Museum of Contemporary Art in Tokyo (2015), the 20th Biennale of Sydney (2015), and the K11 Art Foundation (2020). While towns like Nahara had "re-opened" as early as 2015, artist Meiro Koizumi's site-specific sound installation in Futuba City was the first of the twelve projects to become publicly accessible in October 2022.

FIGURE 15.8 Nikolaus Hirsch and Jorge Otero-Pailos, *Becoming Monument*, 2015–ongoing. Pump house and contents, 7m diameter circle, Fukushima exclusion zone, Japan. Courtesy of the artists and Don't Follow the Wind. HD video, 40 minutes
COURTESY OF THE ARTIST AND VIDEO-FORUM OF NEUER BERLINER KUNSTVEREIN, BERLIN, GERMANY

1970s,[47] *Becoming Monument* measures the lifespan of surface radiation of a pump house and its contents.[48] Within an undisturbed seven-meter-diameter circle, the project implements the same scientific monitoring protocols used on the world's most important monuments, including assessments such as "surface radioactivity, moisture, color variation, structural deformation, and surface cracking."[49] The ongoing monitoring of the building's radioactive dust,

47 Nikolaus Hirsch and Jason Waite, eds., "Don't Follow the Wind," *Critical Spatial Practice* 12 (London: Sternberg Press, 2021), 13.

48 Robert Barry, an American conceptual artist, created a series of artworks using radiation as his medium during this period. For example in January 1969, he created *0.5 Microcurie Radiation Installation* (1969), where he buried four capsules of Barium-133 behind the Metropolitan Museum of Art in New York City. Barium-133, a relatively stable isotope that emits a non-harmful level of charged energy, would continue to discharge a low level of radiation until it reached its approximate half-life of ten years, marking the culmination of the work.

49 Hirsch and Waite, *Don't Follow the Wind*, 130.

an act that blends fieldwork with performance art, will continue for at least 30 years when the unstable isotopes reach their half-life. The process of capturing the dust's evolution over time constitutes the essence of the project, shifting an everyday architecture into a monument for atomic violence and precarious timescales.[50]

7 Conclusion

Returning to the measurement model used to calculate Coordinated Universal Time, the variable uncertainty arising from each stage of the global timekeeping process signals an exponential proliferation of instability. As with timekeeping, uncertainty permeates every stratum of daily life, mirroring the lived realities of late-stage capitalism, climate collapse, global pandemics, and political conflict. This instability feels like it is not merely growing but accelerating and mutating in unpredictable ways.

In his book, *New Dark Age: Technology and the End of the Future*, technologist and artist James Bridle says, "Uncertainty, mathematically and scientifically understood, is not the same as unknowing. Uncertainty, in scientific, climatological terms, is a measure of precisely what we do know. And as our computational systems expand, they show us ever more clearly how much we do not know."[51] As we approach a new era of timekeeping with the advent of quantum lattice clocks—a design intended to erase uncertainty altogether—this innovation only widens the felt "drift" between our daily lived experience and the time-critical computational processes that underpin the always-on rhythms of our globalized world.

We now find ourselves in what philosopher Byung-Chul Han calls an era of *dyschronicity*, reflecting the temporal disjunction that has arisen from the atomization of time, where durational experiences feel detached from

---

50 In addition to radioactive substances released, scientists identified an environmental domino effect after the tsunami, in which thousands of destroyed buildings subsequently released halocarbon gases from damaged refrigerators, air conditioners and insulation. The studies concluded that large-scale catastrophes expedite the release of chemical emissions into the atmosphere and accelerate global warming. As reported by the National Institute for Environmental Studies in Tsukuba, the earthquake-related surge of halocarbons increased ozone loss from Japanese emissions by 38 percent from March 2011 to February 2012.

51 James Bridle, *New Dark Age: Technology and the End of the Future* (New York: Verso, 2018), 44.

FIGURE 15.9 *The Propagation of Uncertainty*, 2020, three-channel video installation, 5 min. 40 sec. Video still
PHOTO © EMILY DICARLO

historical structures of order.[52] As measured duration shifts further into the nano-temporal realm, the computational "black box" tempos become more and more uncoupled from the natural, biological rhythms of time. Instead, they emphasize a recurring presentness of technological operations, where instantaneous processes sustain a non-linear, densely layered temporal structure.

Just as scientists edited out the idiosyncratic, the anomalous and strange from their visual records centuries ago, so do computational systems now. Bridle concludes, "That which is hard to quantify and difficult to model, that which has not been seen before or which does not map onto established patterns, that which is uncertain or ambiguous, is excluded from the field of possible futures. Computation projects a future that is like the past—which makes it, in turn, incapable of dealing with the reality of the present, which is never stable."[53]

If we are living in a temporal crisis as Han suggests, then the role of the artist is to celebrate the unquantifiable, seize the potential of the variable and

52 Byung-Chul Han, *The Scent of Time: A Philosophical Essay on the Art of Lingering* (Cambridge, UK: Polity Press, 2017), iv.

53 Bridle, *New Dark Age*, 44.

through creative intervention, be the *noise* that interrupts the dyschronic density of networked time. A glitch in the tick. A jitter in the signal. Embrace the drift to imagine a future that pulses to a rhythm unbound by standardized time.

References

Barry, Robert. *0.5 Microcurie Radiation Installation*, 1969. Site-specific installation. Metropolitan Museum of Art, New York City.

Brettkelly-Chalmers, Kate. *Time, Duration, and Change in Contemporary Art: Beyond the Clock*. Bristol: Intellect Ltd., 2019.

Bridle, James. *New Dark Age: Technology and the End of the Future*. New York: Verso, 2018.

Burdick, Alan. *Why Time Flies: A Mostly Scientific Investigation*. New York: Simon & Schuster, 2017.

Canales, Jimena. *A Tenth of a Second: A History*. Chicago: The University of Chicago Press, 2009.

Crary, Jonathan. *24/7: Late Capitalism and the Ends of Sleep*. New York: Verso, 2013.

Daston, Lorraine and Peter Galison. *Objectivity*. New York: Zone Books, 2007.

Dahlberg, Kajsa. *Reach, Grasp, Move, Position, Apply Force*. 2015. HD Video. In *Clocking Out: Time Beyond Management*. e-flux Screening Room, New York City, May 20, 2023.

Delfanti, Alessandro. *The Warehouse: Workers and Robots at Amazon*. London: Pluto Press, 2021.

DiCarlo, Emily. *The Propagation of Uncertainty*. 2020. Three-channel video installation. Art Museum, University of Toronto, Canada.

Ernst, Wolfgang. *Chronopoetics: The Temporal Being and Operativity of Technological Media*. Lanhan: Rowman & Littlefield International, 2016.

Gertsvolf, Marina. "Personal interview with the Team Lead of the Time and Frequency Department, National Research Council." Ottawa, Canada, March 30, 2020.

Han, Byung-Chul. *The Scent of Time: A Philosophical Essay on the Art of Lingering*. Cambridge, UK: Polity Press, 2017.

Hannah, Dehlia, and Nadim Samman. "Emily DiCarlo: The Propagation of Uncertainty." *University of Toronto MVS Studio Program Graduating Exhibition*. Toronto: Art Museum, 2020.

Hirsch, Nikolaus and Jason Waite (eds.). "Don't Follow the Wind." *Critical Spatial Practice* 12. London: Sternberg Press, 2021.

James, William. *The Principles of Psychology*. New York: H. Holt, 1890.

Joint Committee for Guides in Metrology (JCGM). "Guide to the Expression of Uncertainty in Measurement—Part 6: Developing and Using Measurement Models." *JCGM GUM-6:2020*, 2020.

Lewandowski, W., D. Matsakis, G. Panfilo, and P. Tavella. "The Evaluation of Uncertainties in [UTC–UTC(k)]." *Metrologia*, vol. 43. no. 3 (2006).

National Institute of Information and Communications Technology. *Japan Standard Time—history*. Accessed April 30, 2024, https://www.nict.go.jp/en/sts/history.html.

National Institute of Information and Communications Technology. Decentralized Japan Standard Time: Time-link between separated facilities. Accessed April 30, 2024. https://www.nict.go.jp/en/sts/decentralized_jst.html.

National Institute of Information and Communications Technology. Report of Time and Frequency Activities at NICT (2012–2015). Accessed June 3, 2021. https://www.bipm.org/documents/20126/28435182/working-document-ID-5264/03fbf334-13b1-9a83-db1d-296919adfbda.

National Institute of Standards and Technology. *SI Redefinition: Meter*. U.S. Department of Commerce, June 2, 2021. Accessed April 11, 2024. https://www.nist.gov/si-redefinition/meter/.

National Metrology Institute of Japan (NMIJ)/AIST, *19th Meeting of CCTF, Research Activities on Time and Frequency*. Accessed March 20, 2020. https://pdfs.semanticscholar.org/4b53/38dc2a9f1c3d54e023ebb5817f677a690a4b.pdf.

National Metrology Institute of Japan (NMIJ) and National Institute of Advanced Industrial Science and Technology (AIST). "The current status of Acoustics, Ultrasound and Vibration measurement standards at NMIJ/AIST." BIPM. Accessed March 20, 2020. https://www.bipm.org/documents/20126/28435028/working-document-ID-5165/97806bc9-7aac-6958-1f5e-96e33c19c0da.

O'Brien, Matt. "Tech disruptions sparked by software update highlight the fragility of globally connected technology." *AP News*, July 19, 2024. Accessed January 10, 2025. https://apnews.com/article/tech-outage-crowdstrike-microsoft-windows-7294e6c9356050cb095dad3353e9244c.

Salame, Richard. "The New Taylorism." *Jacobin*, February 20, 2018. Accessed March 4, 2024. https://jacobin.com/2018/02/amazon-wristband-surveillance-scientific-management.

Scheuregger, Martin. "Personal interview with Senior Lecturer in Music in the School of Creative Arts." University of Lincoln, United Kingdom, November 15, 2023.

Sharma, Sarah. *In the Meantime: Temporality and Cultural Politics*. Durham: Duke University Press, 2014.

The International Bureau of Weights and Measures. *Circular T: no. 382* (November 07, 2019, 15h UTC). France, Paris, 2019. Accessed January 30, 2020. https://www.bipm.org/en/time-ftp/circular-t.

Takala, Pilvi. *The Trainee*, 2008. Installation. Three videos with a duration of 13:52, PowerPoint presentation, keycard, letter. https://pilvitakala.com/the-trainee.

Thorell, Johanna. "Measures of Time: Countering the Clock, Troubling the Timeline." *Clocking Out: Time Beyond Measurement*. New York: Whitney Museum of American Art, 2023.

Tavella, Patrizia. "Personal interview with the Director of the Time Department at the Bureau of Weights and Measures in Paris, France." February 17, 2020.

Yasuda, Masami. "Personal interview with the Group Leader for Time Standards at the Research Institute for Physical Measurement and National Metrology Institute of Japan." Tsukuba, Japan, July 19, 2023.

Index

www.ingramcontent.com/pod-product-compliance
Lightning Source LLC
Chambersburg PA
CBHW071333230926
51849CB00002B/388
9789004723627